GLENCOE MATH

Interactive Guide

CONTRIBUTING AUTHORS

Philip Gonsalves
Director of Curriculum and Instruction for Mathematics
West Contra Costa Unified School District
Richmond, California

Dinah Zike
Educational Consultant
Dinah-Might Activities, Inc.
San Antonio, Texas

McGraw Hill Education

Bothell, WA • Chicago, IL • Columbus, OH • New York, NY

mheducation.com/prek-12

STEM McGraw-Hill is committed to providing
instructional materials in Science, Technology, Engineering,
and Mathematics (STEM) that give all students a solid
foundation, one that prepares them for college and careers
in the 21st century.

Send all inquiries to:
McGraw-Hill Education
STEM Learning Solutions Center
8787 Orion Place
Columbus, OH 43240

ISBN: 978-0-07-898943-8
MHID: 0-07-898943-4

Printed in the United States of America.

1 2 3 4 5 6 7 8 9 LWI 22 21 20 19 18 17

Visual Kinesthetic Vocabulary® is a registered trademark of
Dinah-Might Adventures, LP.

Contents

Chapter 9 Scatter Plots and Data Analysis

Lesson 1 Vocabulary
Rational Numbers

Use the vocabulary squares to write a definition, a sentence, and an example for each vocabulary word.

rational number	Definition
Example	**Sentence**

repeating decimal	Definition
Example	**Sentence**

terminating decimal	Definition
Example	**Sentence**

Lesson 2 Vocabulary

Powers and Exponents

Use the word bank to identify the parts of the expression. Draw an arrow from the word to the part of the expression it describes. Then use the three column chart to organize the vocabulary. Write the word in Spanish. Then write the definition of each word.

Word Bank			
power	base	exponent	factor

$$2^3 = 2 \cdot 2 \cdot 2$$

English	Spanish	Definition
power		
base		
exponent		

Lesson 3 Vocabulary
Multiply and Divide Monomials

Use the definition map to list qualities about the vocabulary word or phrase.

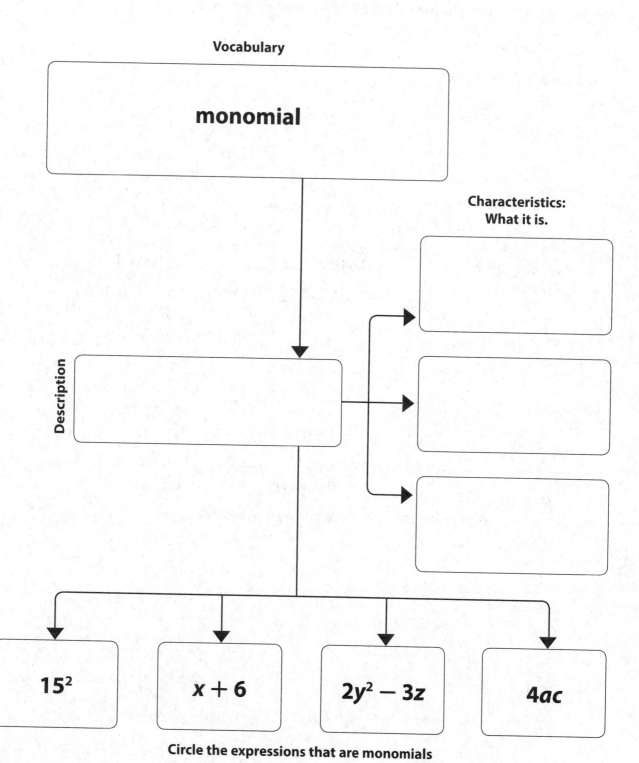

Vocabulary

monomial

**Characteristics:
What it is.**

Description

15^2

$x + 6$

$2y^2 - 3z$

$4ac$

Circle the expressions that are monomials

Lesson 4 Notetaking

Powers of Monomials

Use Cornell notes to better understand the lesson's concepts. Complete each sentence by filling in the blanks with the correct word or phrase.

Questions	Notes
1. How do I find the power of a power?	I can _____ the _____ .
2. How do I find the power of a product?	I can find the _____ of each _____ and _____ .

Summary

How does the Product of Powers law apply to finding the power of a power?

_____

_____

_____

_____

_____

_____

_____

Problem-Solving Investigation
The Four-Step Plan

Case 3 Class Trip

All of Mr. Bassetts' science classes are going to the Natural History Museum.

A tour guide is needed for each **group of eight students**.

His classes have **28** students, **35** students, **22** students, **33** students, **and 22** students.

How many tour guides are needed?

- Understand:

- Plan:

- Solve:

- Check:

Case 4 Gardening

Mrs. Lopez is designing her garden in the shape of a rectangle.

The **area of her garden is 2 times greater than the area of the rectangle shown**.

Write the area of Mrs. Lopez's garden in simplest form.

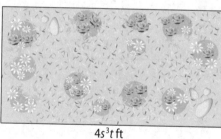

$8s^2$ ft

$4s^3t$ ft

- Understand:

- Plan:

- Solve:

- Check:

Lesson 5 Review Vocabulary
Negative Exponents

Use the definition map to list qualities about the vocabulary word or phrase.

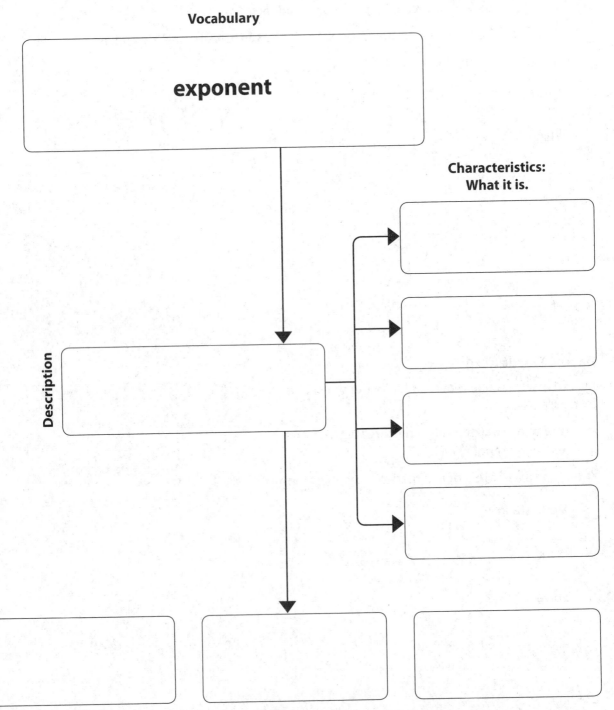

Vocabulary

exponent

**Characteristics:
What it is.**

Description

Write examples of terms that contain a negative or zero exponent.

Lesson 6 Vocabulary
Scientific Notation

Use the vocabulary squares to write a definition, a sentence, and an example for each vocabulary word.

powers of 10	Definition
Example	**When would you use this?**

standard form of a number	Definition
Example	**When would you use this?**

scientific notation	Definition
Example	**When would you use this?**

Lesson 7 Notetaking

Compute with Scientific Notation

Use Cornell notes to better understand the lesson's concepts. Complete each sentence by filling in the blanks with the correct word or phrase.

Questions	Notes
1. How do I multiply and divide with scientific notation?	I can _____ the _____ and use the Product of Powers to _____ the _____ . I can _____ the _____ and use the Quotient of Powers to _____ the _____ .
2. How do I add and subtract with scientific notation?	First, I line up the _____ and rewrite each expression with the same power of _____ . Then, I use the Distributive Property and add or subtract the _____ . Finally, I _____ in scientific form.

Summary
How does scientific notation make it easier to perform computations with very large or very small numbers? _____ _____ _____ _____ _____ _____ _____

Inquiry Lab Guided Writing

Graphing Technology: Scientific Notation Using Technology

WHAT are the similarities and differences between a number written in scientific notation and the calculator notation of the number shown on a screen?

Use the exercises below to help answer the Inquiry Question. Write the correct word or phrase on the lines provided.

1. Rewrite the question in your own words.

   _____

   _____

2. What key words do you see in the question?

   _____

   _____

3. Similarities are ways in which things are _____ .

4. Reread the question. What word is the opposite of *similarities*? _____

5. A number written as the product of a factor and an integer power of ten

   is written in what form? _____

6. Write *scientific notation* or *calculator notation* to label each number below.

   a. 7.23E18   **b.** 7.23×10^{18}

   _____   _____

7. What factor is found in both forms of notation shown above? _____

8. How is the power of ten written in scientific notation? _____

9. How is the power of ten shown in calculator notation?

   _____

WHAT are the similarities and differences between a number written in scientific notation and the calculator notation of the number shown on a screen?

_____

_____

_____

Lesson 8 Vocabulary

Roots

Use the three column chart to organize the vocabulary in this lesson. Write the word in Spanish. Then write the definition of each word.

English	Spanish	Definition
square root		
perfect square		
radical sign		
cube root		
perfect cube		

Inquiry Lab Guided Writing

Roots of Non-Perfect Squares

HOW can you estimate the square root of a non-perfect square number?

Use the exercises below to help answer the Inquiry Question. Write the correct word or phrase on the lines provided.

1. Rewrite the question in your own words.

_____

_____

2. What key words do you see in the question?

_____

3. To _____ means to tell *about* how much instead of finding an exact answer.

4. The square root of 16 is 4. Sixteen is an example of a _____ square number because its square root is a whole number.

5. A number whose square root is not a whole number is

a _____ .

6. What is the next whole number greater than 4? _____

What is its square? _____

7. So, the square root of the whole numbers 17 through 24 would be

between what two numbers? _____

8. Would the square root of 23 be closer to 4 or 5? _____

HOW can you estimate the square root of a non-perfect square number?

_____

_____

_____

Lesson 9 Notetaking

Estimate Roots

Use Cornell notes to better understand the lesson's concepts. Complete each sentence by filling in the blanks with the correct word or phrase.

Questions	Notes
1. How do I estimate a square root?	First, I determine if the square root is a perfect _____ . If not, then I use a _____ to determine between which two perfect _____ the square root falls between and estimate based on where the square root falls on the number line.
2. How do I estimate a cube root?	First, I determine if the cube root is a perfect _____ . If not, then I use a _____ to determine between which two perfect _____ the cube root falls between and estimate based on where the cube root falls on the number line.

Summary
How can I estimate the square root of a non-perfect square?   _____   _____   _____   _____   _____   _____

Lesson 10 Vocabulary
Compare Real Numbers

Use the word cards to define each vocabulary word or phrase and give an example.

Word Cards

irrational number	números irracionales
Definition	**Definición**
_____	_____
_____	_____
_____	_____
Example Sentence	
_____	
_____	

Copyright © McGraw-Hill Education

Word Cards

real numbers	número real
Definition	**Definición**
_____	_____
_____	_____
_____	_____
Example Sentence	
_____	
_____	

Copyright © McGraw-Hill Education

Lesson 1 Vocabulary
Solve Equations with Rational Coefficients

Use the word cards to define each vocabulary word or phrase and give an example.

multiplicative inverse

Definition

_____

_____

_____

Example Sentence

_____

_____

inversos multiplicativo

Definición

_____

_____

_____

- -

coefficient

Definition

_____

_____

_____

Example Sentence

_____

_____

coeficiente

Definición

_____

_____

_____

Inquiry Lab Guided Writing

Solve Two-Step Equations

HOW does a bar diagram help you solve a real-world problem involving a two-step equation?

Use the exercises below to help answer the Inquiry Question. Write the correct word or phrase on the lines provided.

1. Rewrite the question in your own words.

_____

_____

2. What key words do you see in the question?

_____

3. You can use a _____ diagram to model problems.

4. A bar diagram can be separated into parts. What does the full length of

a bar diagram represent? _____

Use the information below to label the bar diagram in Exercise 5. Then use the bar diagram to answer Exercises 6 and 7.

Five brothers earn a total of $62 for doing chores.

Two brothers earn $10 each.

The other three brothers each earn the same amount.

How much does each of the other three brothers earn?

5.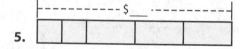

6. How does the bar diagram show that there are 5 brothers?

_____

7. What problem-solving strategy will helps you find the unknown parts? _____

HOW does a bar diagram help you solve a real-world problem involving a two-step equation?

_____

_____

_____

Lesson 2 Vocabulary

Solve Two-Step Equations

**Use the three column chart to organize the vocabulary in this lesson.
Write the word in Spanish. Then write the definition of each word.**

English	Spanish	Definition
properties		
Addition Property of Equality		
Subtraction Property of Equality		
Division Property of Equality		
Multiplication Property of Equality		
two-step equation		

Lesson 3 Review Vocabulary

Write Two-Step Equations

Use the flow chart to review the process for translating sentences into equations.

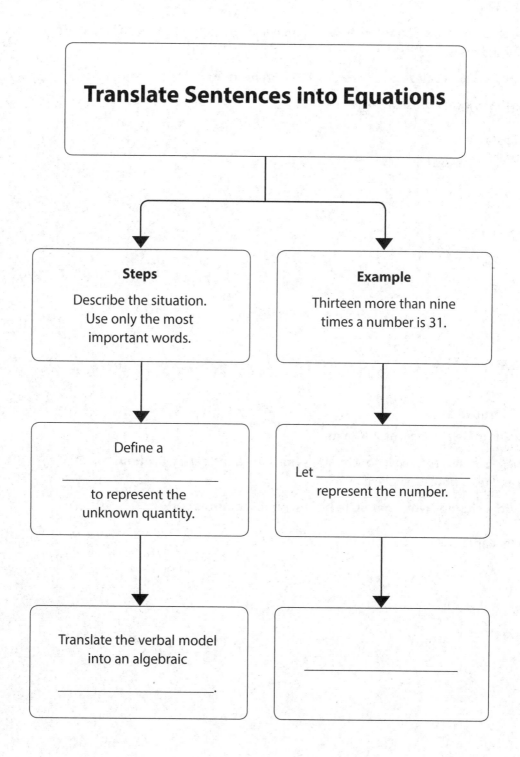

Translate Sentences into Equations

Steps

Describe the situation. Use only the most important words.

Example

Thirteen more than nine times a number is 31.

Define a

_____

to represent the unknown quantity.

Let _____

represent the number.

Translate the verbal model into an algebraic

_____.

_____

Problem-Solving Investigation
Work Backward

Case 3 Financial Literacy

Janelle has **$75**.

She buys jeans that are **discounted 30%** and then uses an in-store coupon for **$10 off** the discounted price.

After paying **$3.26 in sales tax**, she receives **$17.34 in change**.

What was the **original price** of the jeans?

- Understand:

- Plan:

- Solve:

- Check:

Case 4 Schedule

Nyoko needs to be at school at **7:45 a.m.**

It takes her **15 minutes** to walk to school, $\frac{5}{12}$ **hour** to eat breakfast, **0.7 hour** to get dressed, and **0.15 hour** to shower.

At what time should Nyoko get up to be at school **5 minutes early**?

- Understand:

- Plan:

- Solve:

- Check:

Inquiry Lab Guided Writing

Equations with Variables on Each Side

**HOW do you use the Properties of Equality when solving
an equation using algebra tiles?**

**Use the exercises below to help answer the Inquiry Question. Write the correct
word or phrase on the lines provided.**

1. Rewrite the question in your own words.

_____

_____

2. What key words do you see in the question?

_____

3. Algebra tiles are used to _____.

4. The Properties of Equality tell us to make the same changes on each side of the

_____ sign in an equation.

5. With what operations can the Properties of Equality be used?

_____

Model the equation $8x - 9 = 4x + 3$ with algebra tiles to answer Exercises 6-10.

6. How many x-tiles can you subtract from each side of the equation? _____

7. Write the resulting equation. _____

8. How many 1-tiles can you add to each side of the equation? _____

9. Write the resulting equation. _____

10. Into how many groups will you divide the 1-tiles to find the value of x? _____

What is the solution? _____

HOW do you use the Properties of Equality when solving an equation using algebra tiles?

_____

_____

_____

Lesson 4 Review Vocabulary

Solve Equations with Variables on Each Side

Use the definition map to list qualities about the vocabulary word or phrase.

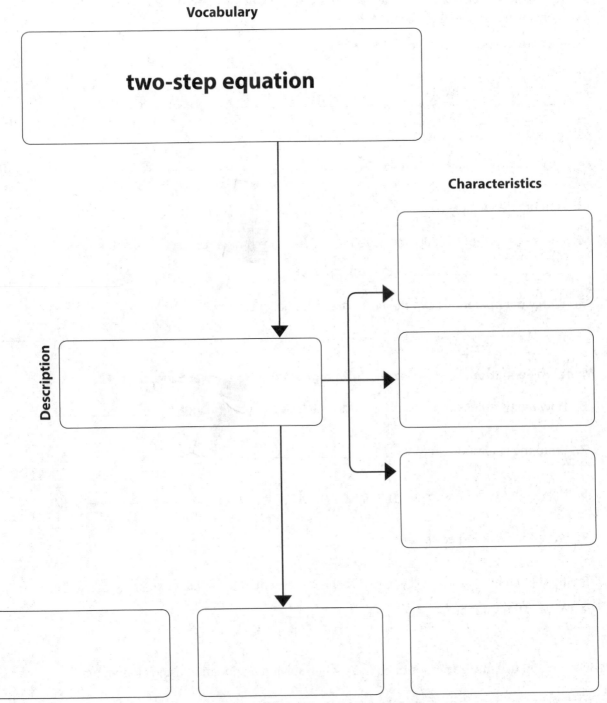

Vocabulary

two-step equation

Characteristics

Description

Write three examples of two-step equations.

Lesson 5 Vocabulary
Solve Multi-Step Equations

Use the word cards to define each vocabulary word or phrase and give an example.

Word Cards

null set	**conjunto nulo**
Definition	**Definición**
_____	_____
_____	_____
_____	_____
Example Sentence	
_____	
_____	

Word Cards

identity	**identidad**
Definition	**Definición**
_____	_____
_____	_____
_____	_____
Example Sentence	
_____	
_____	

Lesson 1 Vocabulary

Constant Rate of Change

Use the word cards to define each vocabulary word or phrase and give an example.

Word Cards

linear relationship

Definition

_____

_____

_____

Example Sentence

_____

_____

relación lineal

Definición

_____

_____

_____

Word Cards

constant rate of change

Definition

_____

_____

_____

Example Sentence

_____

_____

tasa constant de cambio

Definición

_____

_____

_____

Inquiry Lab Guided Writing
Graphing Technology: Rate of Change

HOW can you use a graphing calculator to determine the rate of change?

Use the exercises below to help answer the Inquiry Question. Write the correct word or phrase on the lines provided.

1. Rewrite the question in your own words.

   _____

   _____

2. What key words do you see in the question?

   _____

3. _____ of change describes how one quantity changes in relation to another.

4. A linear relationship has a _____ rate of change.

   Its graph is a straight _____ .

5. An example of an _____ that can be graphed is $y = 6x$.

6. Suppose you graph the equation in Exercise 5 on a graphing calculator.

   What feature can you use to find two points on the line? _____

7. You can use any two points on the line to find the rate of _____ .

8. Which lines are steeper, those with a greater rate of change or

   a lesser rate of change? _____

HOW can you use a graphing calculator to determine the rate of change?

_____

_____

_____

Lesson 2 Vocabulary

Slope

Use the concept web to define slope in five different ways.

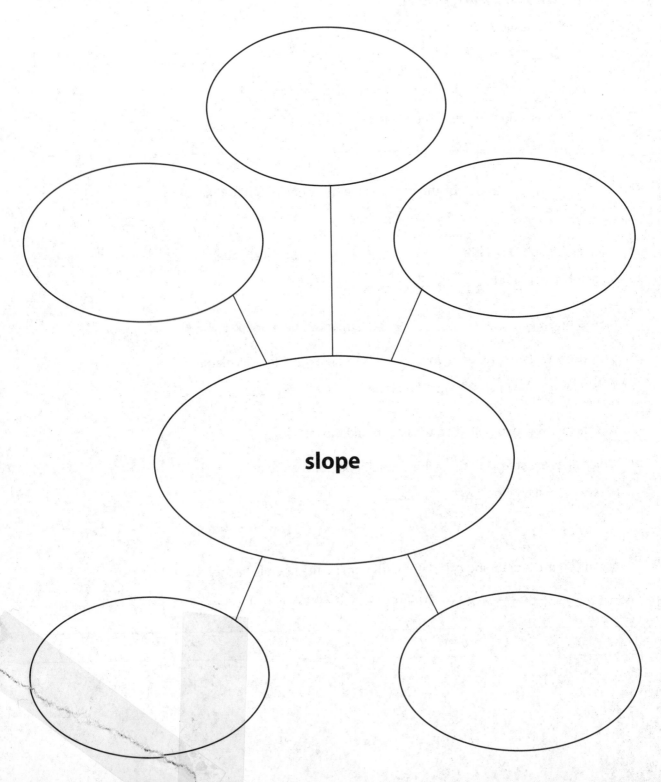

Lesson 3 Vocabulary

Equations in y = mx Form

Use the vocabulary squares to write a definition, a sentence, and an example for each vocabulary word.

	Definition
direct variation	
Example	**Sentence**

	Definition
constant of variation	
Example	**Sentence**

	Definition
constant of proportionality	
Example	**Sentence**

Lesson 4 Vocabulary

Slope-Intercept Form

Use the concept web and the word bank to identify the parts of an equation in slope-intercept form.

Word Bank			
slope	*x*-coordinate	*y*-coordinate	*y*-intercept

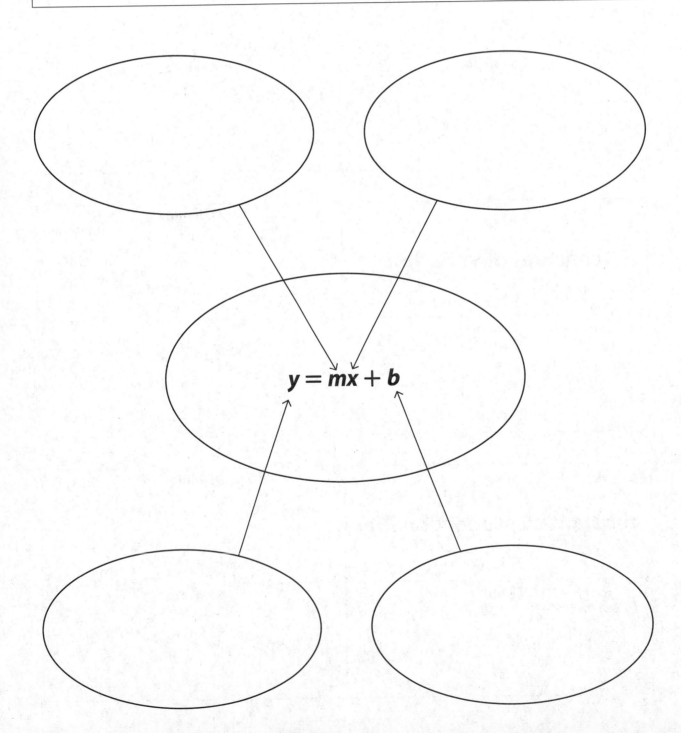

$$y = mx + b$$

Inquiry Lab Guided Writing

Slope Triangles

HOW does graphing slope triangles on the coordinate plane help you analyze them?

Use the exercises below to help answer the Inquiry Question. Write the correct word or phrase on the lines provided.

1. Rewrite the question in your own words.

_____

_____

2. What key words do you see in the question?

_____

3. The vertical change in a line is called _____ .

4. The horizontal change in a line is called _____ .

5. _____ is the ratio of the rise to the run.

Use the slope triangles below to answer Exercises 6-8.

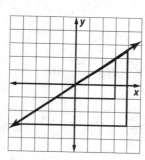

6. Are the angle measures equal in the two triangles? _____

7. Do the triangles have the same shape? _____

8. Are the triangles the same size? _____

HOW does graphing slope triangles on the coordinate plane help you analyze them?

_____

_____

_____

Lesson 5 Vocabulary
Graph a Line Using Intercepts

Use the word cards to define each vocabulary word or phrase and give an example.

Word Cards

x-intercept

Definition

_____

_____

_____

Example Sentence

_____

_____

intersección x

Definición

_____

_____

_____

Word Cards

standard form

Definition

_____

_____

_____

Example Sentence

_____

_____

forma estándar

Definición

_____

_____

_____

Problem-Solving Investigation
Guess, Check, and Revise

Case 3 Wrap it Up

Shya works part-time at a gift-wrapping store.

The store sells wrapping paper rolls and square packages of wrapping paper.

There are a total of **125 rolls and packages.**

Each roll costs $3.50 and each package costs $2.25.

The **total cost** of all of the rolls and packages is **$347.50.**

How many rolls of wrapping paper are there?

- Understand:

- Plan:

- Solve:

- Check:

Case 4 Family

Five siblings have a **combined age of 195 years.**

The **oldest is 13 years older than the youngest,** Marc.

The **middle child,** Anne, **is five years younger than Josie.**

The **other two** siblings **are 6 years apart.**

If the **second oldest child is 42,** what are the **ages of the siblings?**

- Understand:

- Plan:

- Solve:

- Check:

Lesson 6 Review Vocabulary

Write Linear Equations

The items in the word bank represent given information. Match the correct item to the process you would use to write a linear equation.

Word Bank		
slope & one point		slope & y-intercept
graph	two points	table

	Use the coordinates to find the slope. Substitute the slope and coordinates of one of the points in $y - y_1 = m(x - x_1)$.
	Find the y-intercept b and the slope m, and then substitute the slope and y-intercept b in $y = mx + b$.
	Substitute the slope m and y-intercept b in $y = mx + b$.
	Substitute the slope m and coordinates in $y - y_1 = m(x - x_1)$.
	Find two points and calculate the slope. Then substitute the slope and coordinates of one of the points in $y - y_1 = m(x - x_1)$.

Inquiry Lab Guided Writing

Graphing Technology: Model Linear Behavior

HOW does using technology help you to determine if situations display linear behavior?

Use the exercises below to help answer the Inquiry Question. Write the correct word or phrase on the lines provided.

1. Rewrite the question in your own words.

_____

_____

2. What key words do you see in the question?

_____

Use the graphs below to answer Exercises 3-6.

A

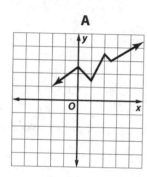

B

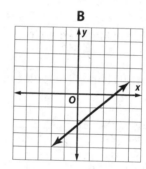

3. Which graph displays linear behavior? _____

4. Which graph does not have a constant rate of change? _____

5. In the Inquiry Lab, what tool is used to graph a situation? _____

6. The graph of linear behavior is a _____ line.

HOW does using technology help you to determine if situations display linear behavior?

_____

_____

_____

Inquiry Lab Guided Writing

Graphing Technology: Systems of Equations

HOW can I use a graphing calculator to find one solution for a set of two equations?

Use the exercises below to help answer the Inquiry Question. Write the correct word or phrase on the lines provided.

1. Rewrite the question in your own words.

_____

_____

2. What key words do you see in the question?

_____

3. What button do you use to graph equations on a calculator? _____

4. How many lines will the calculator show if you enter two equations? _____

5. Where the two lines cross is called the point of _____ .

6. The point of intersection is the _____ of the set of two equations.

7. Write the solution shown on the graph below. _____

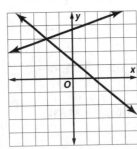

HOW can I use a graphing calculator to find one solution for a set of two equations?

_____

_____

_____

Lesson 7 Vocabulary

Solve Systems of Equations by Graphing

Use the definition map to list qualities about the vocabulary word or phrase.

Vocabulary

> ## system of equations

Characteristics of the solution

Description

Examples

Lesson 8 Vocabulary

Solve Systems of Equations Algebraically

Use the word cards to define each vocabulary word or phrase and give an example.

Word Cards

substitution	sustitución
Definition	**Definición**
_____	_____
_____	_____
_____	_____
Example Sentence	
_____	
_____	

- -

Word Cards

variable	variable
Definition	**Definición**
_____	_____
_____	_____
_____	_____
Example Sentence	
_____	
_____	

Inquiry Lab Guided Writing

Analyze Systems of Equations

HOW can you solve real-world mathematical problems using two linear equations in two variables?

Use the exercises below to help answer the Inquiry Question. Write the correct word or phrase on the lines provided.

1. Rewrite the question in your own words.

_____

_____

2. What key words do you see in the question?

_____

_____

3. In the Inquiry Lab, you drew _____ segments on a graph to represent two equations.

4. The lines will intersect at one point if the _____ are not the same.

5. The point of _____ is the solution of the equations.

6. You can also use algebra to solve by writing the equations

as a _____ .

7. A system of equations is two or more equations with the same

set of _____ .

For Exercises 8 and 9, write yes or no to tell whether each pair of equations is a system.

8. $y = 2x + 4$
$y = 4b$

_____

9. $y = x - 8$
$y = -4$

_____

HOW can you solve real-world mathematical problems using two linear equations in two variables?

_____

_____

_____

Lesson 1 Vocabulary

Represent Relationships

Use the definition map to list qualities about the vocabulary word or phrase.

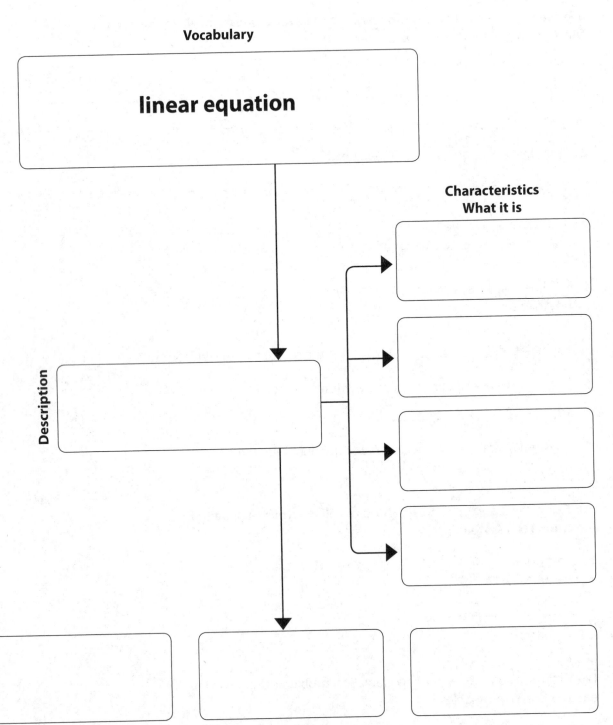

Vocabulary

linear equation

Characteristics
What it is

Description

Write examples of linear equations.

Lesson 2 Vocabulary
Relations

Use the vocabulary squares to write a definition, a sentence, and an example for each vocabulary word.

relation	Definition
Example	**Sentence**

domain	Definition
Example	**Sentence**

range	Definition
Example	**Sentence**

Inquiry Lab Guided Writing

Relations and Functions

HOW can I determine if a relation is a function?

Use the exercises below to help answer the Inquiry Question. Write the correct word or phrase on the lines provided.

1. Rewrite the question in your own words.

   _____

   _____

2. What key words do you see in the question?

   _____

3. Any set of ordered pairs is a _____ .

4. A _____ is a relationship in which there is exactly
   one output value for each input value.

5. Domain is the set of _____ values for a function.

6. Range is the set of _____ values for a function.

Use the mapping diagram for Exercises 7 and 8.

7. Is each member of the domain paired with exactly
   one member of the range? _____

8. Is the relation a function? _____

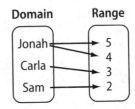

Use the mapping diagram for Exercises 9 and 10.

9. Is each member of the domain paired with exactly
   one member of the range? _____

10. Is the relation a function? _____

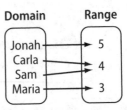

HOW can I determine if a relation is a function?

_____

_____

Lesson 3 Vocabulary

Functions

Use the concept web to describe the different parts of a function table.

Word Bank		
dependent variable	domain	function
independent variable	range	

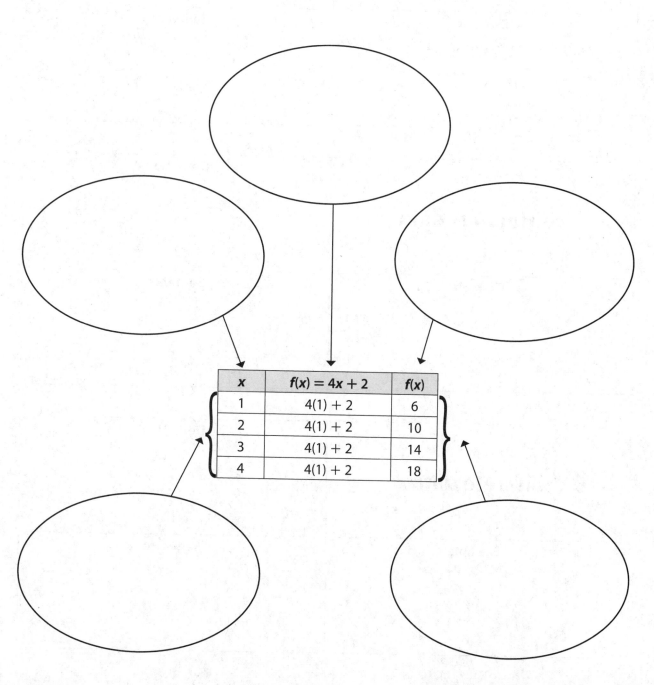

x	$f(x) = 4x + 2$	$f(x)$
1	$4(1) + 2$	6
2	$4(1) + 2$	10
3	$4(1) + 2$	14
4	$4(1) + 2$	18

Lesson 4 Vocabulary

Linear Functions

Use the vocabulary squares to write a definition, a sentence, and an example for each vocabulary word.

	Definition
linear function	
Example	**Sentence**

	Definition
continuous data	
Example	**Sentence**

	Definition
discrete data	
Example	**Sentence**

Problem-Solving Investigation
Make a Table

Case 3 Plants

The table shows the height of a giant bamboo plant for Days 5-9.

The bamboo **grew at a steady rate** each day, starting on Day 5.

From Day 5 to Day 9, the bamboo plant **grew about what percent of its final height**?

Bamboo Growth	
Number of Days	Total Growth (ft)
5	?
6	?
7	10
8	13.5
9	17

- Understand:

- Plan:

- Solve:

- Check:

Case 4 Financial Literacy

Sophia and Scott each open a bank account with **an initial deposit of $50 each**.

Scott planned to save **30% of his earnings** from his after-school job.

The job pays **$8 per hour**, and he works **25 hours each week**.

For **four weeks, Sophia saved $45 per week**.

After that, she saved an additional $30 per week.

During **which week will have they have the same amount of money** in their bank accounts?

- Understand:

- Plan:

- Solve:

- Check:

Lesson 5 Notetaking

Compare Properties of Functions

Use the concept web to show representations of the function in different ways.
Use a graph in one of the pieces of the web.

Equation

Table

x	y

A store sells tuna salad for $4 a pound. They will deliver any size order for a $2 delivery fee.

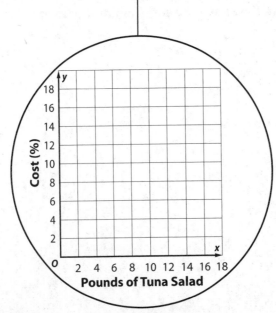

Pounds of Tuna Salad

Lesson 6 Notetaking

Construct Functions

Use Cornell notes to better understand the lesson's concepts. Complete each sentence by filling in the blanks with the correct word or phrase.

Questions	Notes
1. What is the initial value of a function?	The initial value of a function is the corresponding _____ when _____ equals _____ .
2. How do I find the initial value of a function?	I find the _____ or I find the value of _____ when _____ .

Summary

How is the initial value of a function represented in a table and in a graph?

_____

_____

_____

_____

_____

_____

_____

Lesson 7 Vocabulary

Linear and Nonlinear Functions

Use the definition map to list qualities about the vocabulary word or phrase.

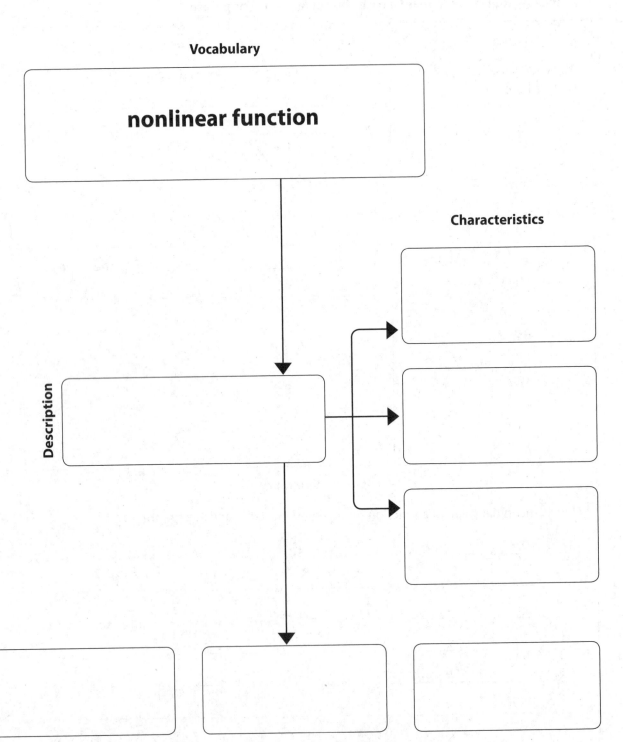

Vocabulary

nonlinear function

Characteristics

Description

Write equations of nonlinear functions.

Lesson 8 Notetaking

Quadratic Functions

Use Cornell notes to better understand the lesson's concepts. Complete each sentence by filling in the blanks with the correct word or phrase.

Questions	Notes
1. What is a quadratic function?	A function in which the greatest _____ of the _____ is _____ ; A quadratic function can be written in the form $y =$ _____ .
2. How do I know if the graph of a quadratic function opens upward or downward?	The graph opens upward if the coefficient of the squared variable is _____. The graph opens downward if the coefficient of the squared variable is _____ .

Summary
When does the graph of a quadratic function open upward or downward?
_____
_____
_____
_____
_____
_____
_____
_____

Inquiry Lab Guided Writing

Graphing Technology: Families of Non-Linear Functions

HOW are families of nonlinear functions the same as the parent function? How are families of nonlinear functions different from the parent function?

Use the exercises below to help answer the Inquiry Question. Write the correct word or phrase on the lines provided.

1. Rewrite the question in your own words.

_____

_____

2. What key words do you see in the question?

_____

3. A function for which the graph is not a straight line is a _____ function.

Use the graph of a family of nonlinear functions below to answer Exercises 4-7.

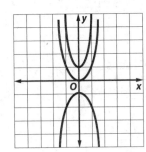

4. Are the shapes of all of the graphs the same? _____

5. Does the location on the axes stay the same from one graph to another? _____

6. Do all graphs have the same width? _____

7. Do all graphs have the same orientation? _____

HOW are families of nonlinear functions the same as the parent function?
How are families of nonlinear functions different from the parent function?

_____

_____

_____

Lesson 9 Vocabulary

Qualitative Graphs

Use the concept web to list characteristics of qualitative graphs. Draw a graph in one of the pieces of the web.

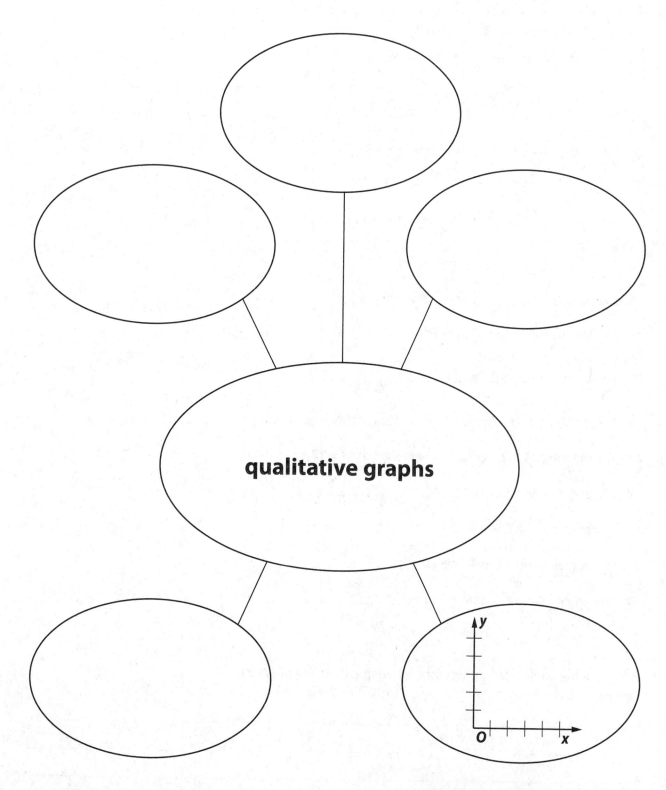

Inquiry Lab Guided Writing

Parallel Lines

WHAT are the angle relationships formed when a third line intersects two parallel lines?

Use the exercises below to help answer the Inquiry Question. Write the correct word or phrase on the lines provided.

1. Rewrite the question in your own words.

_____

_____

2. What key words do you see in the question?

_____

3. Lines in the same plane that never intersect are _____ lines.

4. _____ means "cross."

5. Draw two parallel lines in the space provided.

6. Draw a line that intersects the two parallel lines.

7. How many angles are formed? _____

8. Two angles are supplementary if the sum of their measures is _____ .

Use the drawing at the right to answer Exercises 9-12.

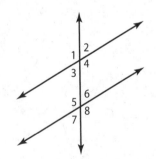

9. Are angles 1 and 5 equal? _____

10. Are angles 5 and 6 equal? _____

11. Are angles 2 and 3 supplementary? _____

12. Are angles 6 and 8 supplementary? _____

WHAT are the angle relationships formed when a third line intersects two parallel lines?

_____

_____

Lesson 1 Vocabulary

Lines

Use the three column chart to write the vocabulary word and definition for each drawing.

What I See	Vocabulary Word	Definition

Lesson 2 Vocabulary

Geometric Proof

Use the flow chart to review the proof process.

> ### Use inductive and/or deductive reasoning to write a proof.

Define inductive reasoning.	Define proof.	Define deductive reasoning.
_____ _____ _____ _____	_____ _____ _____ _____	_____ _____ _____ _____

> ### Write a paragraph proof or a two-column proof using theorems.

Define paragraph or informal proof.	Define theorem.	Define two-column or formal proof.
_____ _____ _____ _____ _____	_____ _____ _____ _____ _____	_____ _____ _____ _____ _____

Inquiry Lab Guided Writing

Triangles

WHAT is the relationship among the measures of the angles of a triangle?

Use the exercises below to help answer the Inquiry Question. Write the correct word or phrase on the lines provided.

1. Rewrite the question in your own words.

_____

_____

2. What key words do you see in the question?

_____

3. How many angles does a triangle have? _____

Use the figure below to answer Exercises 4-8.

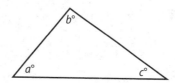

4. If $a = 50°$, $b = 88°$, and $c = 42°$, what is the sum of the angles? _____

5. Use the symbols *a, b,* and *c* to write a formula for the sum of the measures of the angles of a triangle. _____

6. If $a = 70°$ and $b = 65°$, what is the measure of *c*? _____

7. If $b = 83°$ and $c = 39°$, what is the measure of *a*? _____

8. Is the sum of the measures of the angles of a triangle always 180°? _____

WHAT is the relationship among the measures of the angles of a triangle?

_____

_____

_____

Lesson 3 Vocabulary

Angles of Triangles

Use the vocabulary squares to write a definition and a sentence. Then label the figure with an example for each vocabulary word.

interior angle	Definition
	Sentence

exterior angle	Definition
	Sentence

remote interior angles	Definition
	Sentence

Lesson 4 Vocabulary

Polygons and Angles

Use the definition map to list qualities about the vocabulary word or phrase.

Vocabulary

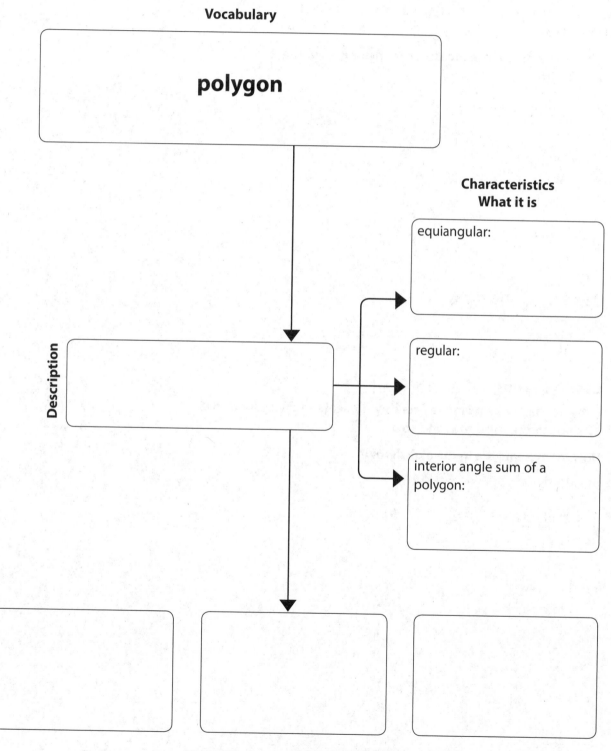

polygon

Description

**Characteristics
What it is**

equiangular:

regular:

interior angle sum of a polygon:

Draw three examples of polygons.

Problem-Solving Investigation
Look for a Pattern

Case 3 Geometry

Right triangles are arranged as shown.

The **sum of the measures of the angles in the first figure is 360°.**

What is the **sum of the measures of the angles in the fifth figure?**

  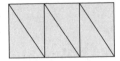

- Understand:

- Plan:

- Solve:

- Check:

Case 4 Seating

A theater has **12 seats in the first row, 17 seats in the second row, 22 seats in the third row, and so on.**

How many seats are in the eighth row?

the **nth row?**

- Understand:

- Plan:

- Solve:

- Check:

Inquiry Lab Guided Writing

Right Triangle Relationships

WHAT is the relationship among the sides of a right triangle?

Use the exercises below to help answer the Inquiry Question. Write the correct word or phrase on the lines provided.

1. Rewrite the question in your own words.

_____

_____

2. What key words do you see in the question?

_____

3. A _____ triangle has one 90° angle.

Use the triangle below to answer Exercises 4-8.

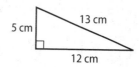

4. What are the measures of the two shortest sides? _____

5. What is the sum of $5^2 + 12^2$? _____

6. What is the measure of the longest side? _____

7. What is 13^2? _____

8. Does the sum of the squares of the two shortest sides equal the

square of the longest side? _____

WHAT is the relationship among the sides of a right triangle?

_____

_____

Lesson 5 Vocabulary

The Pythagorean Theorem

Use the vocabulary squares to write a definition, a sentence, and an example for each vocabulary word.

legs	Definition
Example	**Sentence**

hypotenuse	Definition
Example	**Sentence**

Pythagorean Theorem	Definition
Example	**Sentence**

Inquiry Lab Guided Writing

Proofs About the Pythagorean Theorem

HOW can you prove the Pythagorean Theorem and its converse?

Use the exercises below to help answer the Inquiry Question. Write the correct word or phrase on the lines provided.

1. Rewrite the question in your own words.

_____

_____

2. What key words do you see in the question?

_____

3. Write the Pythagorean Theorem: _____

4. If a triangle has side lengths a, b, and c and $a^2 + b^2 = c^2$, then the triangle is a

_____ triangle.

Use the triangles below to answer Exercises 5 and 6.

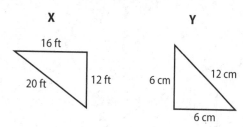

X

Y

16 ft

20 ft

12 ft

6 cm

12 cm

6 cm

5. Which triangle is a right triangle? _____

How do you know? _____

6. Which triangle is not a right triangle? _____

How do you know? _____

7. How does a physical model help you solve problems?

_____

HOW can you prove the Pythagorean Theorem and its converse?

_____

_____

Lesson 6 Notetaking

Use the Pythagorean Theorem

Use vocabulary words and the Pythagorean Theorem to identify the parts and side lengths of the right triangle.

Word Bank		
leg		hypotenuse
$a = \sqrt{c^2 - b^2}$	$b = \sqrt{c^2 - a^2}$	$c = \sqrt{a^2 + b^2}$

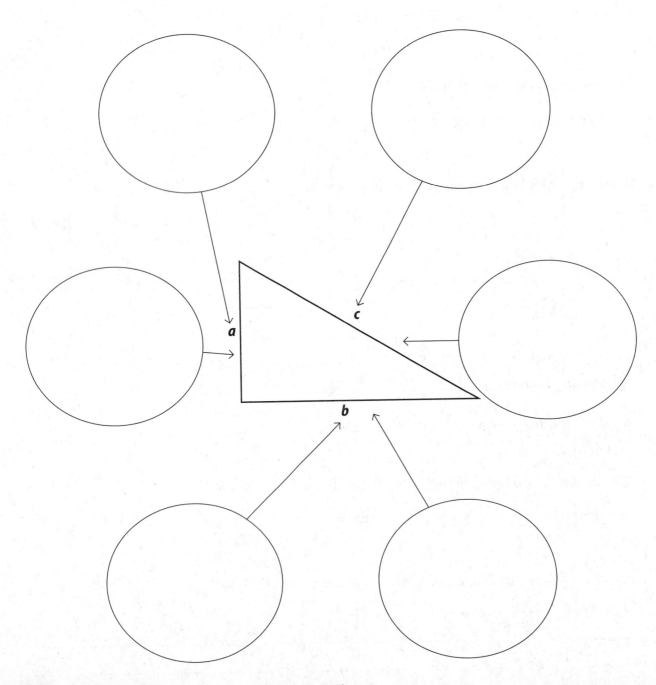

Lesson 7 Notetaking

Distance on the Coordinate Plane

Use the flow chart to review the processes for finding the distance between two points on a coordinate plane.

Use the Pythagorean Theorem or Distance Formula to find the distance between two points on a coordinate plane.

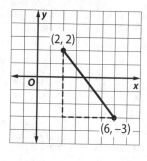

State the Pythagorean Theorem.

_____

State the Distance Formula.

_____

Find the length of the segment using the Pythagorean Theorem.

_____

_____

_____

_____

Find the length of the segment using the Distance Formula.

_____

_____

_____

_____

Inquiry Lab Guided Writing

Transformations

WHAT are some rigid motions of the plane?

Use the exercises below to help answer the Inquiry Question. Write the correct word or phrase on the lines provided.

1. Rewrite the question in your own words.

_____

_____

2. What key words do you see in the question?

_____

3. In a _____ motion of the plane, the shape and size of a figure do not change.

Use the figures at the right to answer Exercises 4 and 5.

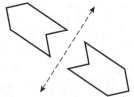

4. Do the figures show a flip, slide, or turn? _____

5. Does the shape or size of the figure change? _____

Use the figures at the right to answer Exercises 6 and 7.

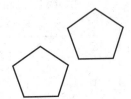

6. Do the figures show a flip, slide, or turn? _____

7. Does the shape or size of the figure change? _____

Use the figures at the right to answer Exercises 8 and 9.

8. Do the figures below show a flip, slide, or turn? _____

9. Does the shape or size of the figure change? _____

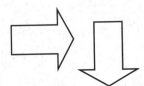

WHAT are some rigid motions of the plane?

_____

_____

_____

Lesson 1 Vocabulary

Translations

Use the three column chart to organize the vocabulary in this lesson. Write the word in Spanish. Then write the definition of each word.

English	Spanish	Definition
transformation		
preimage		
image		
translation		
congruent		

Lesson 2 Vocabulary

Reflections

Use the word cards to define each vocabulary word or phrase and give an example.

Word Cards

reflection	reflexión
Definition	**Definición**
_____	_____
_____	_____
_____	_____
Example Sentence	
_____	
_____	

Word Cards

line of reflection	línea de reflexión
Definition	**Definición**
_____	_____
_____	_____
_____	_____
Example Sentence	
_____	
_____	

Problem-Solving Investigation
Act It Out

Case 3 Picture Exchange

The French Club took a field trip to an exhibit of French art at the museum.

Five of the club **members** held a **picture exchange** to share their pictures.

April brought more pictures **than Brandon.**

Chloe brought more pictures **than Diego, but fewer than Brandon.**

Ethan brought more pictures **than Chloe, but not as many as Brandon.**

List the picture exchange **participants in order from the most pictures to the fewest.**

- Understand:

- Plan:

- Solve:

- Check:

Case 4 Dancing in the Street

In a certain dance for a competition, a dancer makes the following series of steps:
2 steps back, 1 step to the right, 3 steps forward, 2 steps to the right.

The series is **repeated four times.**

How does the dancer's **final position compare to** his **original position**?

- Understand:

- Plan:

- Solve:

- Check:

Inquiry Lab Guided Writing

Rotational Symmetry

HOW can you identify rotational symmetry?

Use the exercises below to help answer the Inquiry Question. Write the correct word or phrase on the lines provided.

1. Rewrite the question in your own words.

_____

_____

2. What key words do you see in the question?

_____

3. Write a synonym for the word rotate. _____

4. If a figure is turned less than 360° on its center point and looks exactly

like the original position, it has _____ symmetry.

Use the figures below to answer Exercises 5-7.

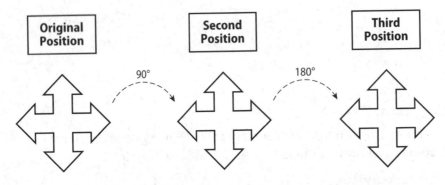

5. By how many degrees was the second figure rotated? _____

Does the second figure look exactly like the orginal figure? _____

6. By how many degrees was the second figure rotated? _____

Does the third figure look exactly like the orginal figure? _____

7. Does the figure have rotational symmetry? _____

HOW can you identify rotational symmetry?

_____

_____

_____

Lesson 3 Vocabulary

Rotations

Use the concept web to name the transformation and the parts of the transformation.

Word Bank	
center of rotation	image
preimage	rotation

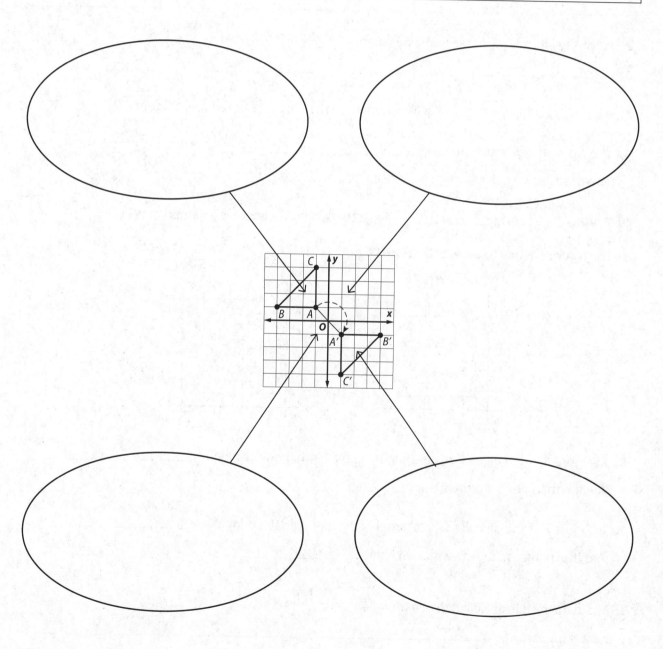

What is the angle of rotation shown in the graph? _____

Inquiry Lab Guided Writing

Dilations

WHAT are the results of a dilation of a triangle?

Use the exercises below to help answer the Inquiry Question. Write the correct word or phrase on the lines provided.

1. Rewrite the question in your own words.

   _____

   _____

2. What key words do you see in the question?

   _____

3. A _____ is a transformation that enlarges or reduces a figure.

4. The word *enlarge* means to make _____ .

   The word *reduce* means to make _____ .

5. A scale _____ is the factor by which a figure is enlarged or reduced.

Use the figures below to answer Exercises 6 and 7.

A

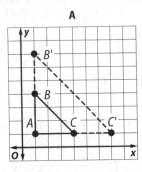

B

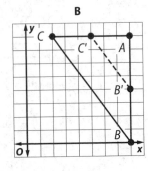

6. In Figure A, is the dilated triangle larger or smaller than the original? _____

   Are the triangles the same shape? _____

7. In Figure B, is the dilated triangle larger or smaller than the orignal? _____

   Are the triangles the same shape in Figure B? _____

   WHAT are the results of a dilation of a triangle?

   _____

   _____

   _____

Lesson 4 Review Vocabulary
Dilations

Use the definition map to list qualities about the vocabulary word or phrase.

Vocabulary

dilation

Describe what happens when
$k > 1$, $k < 1$, and $k = 1$.

Description

Draw and label examples for $k > 1$, $k < 1$, and $k = 1$.

Inquiry Lab Guided Writing

Composition of Transformations

HOW does a combination of transformations differ from a single transformation? How are they the same?

Use the exercises below to help answer the Inquiry Question. Write the correct word or phrase on the lines provided.

1. Rewrite the question in your own words.

_____

_____

2. What key words do you see in the question?

_____

3. When more than one transformation is applied to figure, it is called a

composition of _____ .

Use the transformations below to answer Exercises 4 -6.

Original Image

First Transformation

Second Transformation

4. What is the first transformation? _____

What is the second transformation? _____

5. Can the original image look like the last image with only one transformation? _____

6. Are all of the images the same shape and size? _____

HOW does a combination of transformations differ from a single transformation? How are they the same?

_____

_____

_____

Lesson 1 Review Vocabulary
Congruence and Transformations

Use the definition map to list qualities about the vocabulary word or phrase.

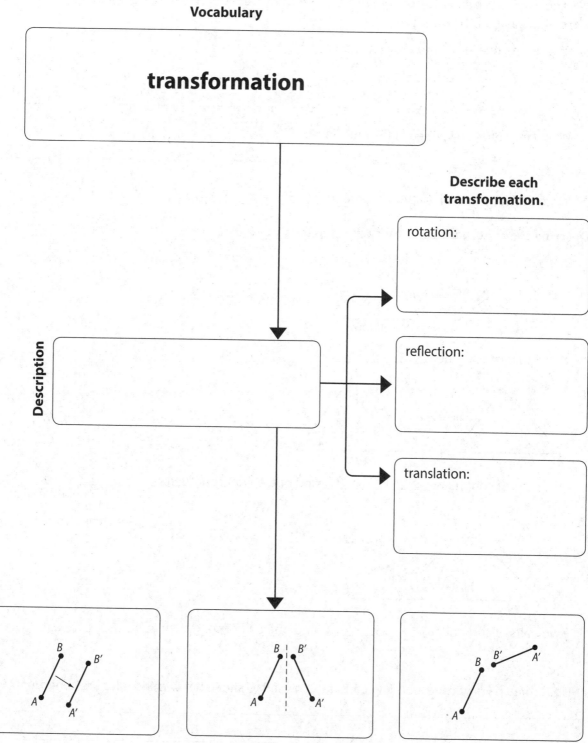

Vocabulary

transformation

Description

Describe each
transformation.

rotation:

reflection:

translation:

Identify each transformation.

Inquiry Lab Guided Writing

Investigate Congruent Triangles

WHICH three pairs of corresponding parts can be used to show that two triangles are congruent?

Use the exercises below to help answer the Inquiry Question. Write the correct word or phrase on the lines provided.

1. Rewrite the question in your own words.

_____

_____

2. What key words do you see in the question?

_____

3. Write a synonym for the word congruent. _____

Describe the congruent parts for each set of triangles.

4.

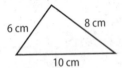 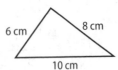

three pairs of congruent _____

5.

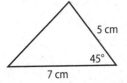 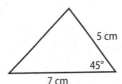

two pairs of congruent _____ and one pair of congruent _____

6.

two pairs of congruent _____ and one pair of congruent _____

WHICH three pairs of corresponding parts can be used to show that two triangles are congruent?

_____

_____

Lesson 2 Vocabulary
Congruence

Use the word cards to define each vocabulary word or phrase and give an example.

```
                          Word Cards
```

congruent	congruente
Definition	**Definición**
_____	_____
_____	_____
_____	_____
Example Sentence	
_____	
_____	

```
                          Word Cards
```

corresponding parts	partes correspondientes
Definition	**Definición**
_____	_____
_____	_____
_____	_____
Example Sentence	
_____	
_____	

Inquiry Lab Guided Writing

Geometry Software

HOW can technology help you show the relationship between transformations and congruence?

Use the exercises below to help answer the Inquiry Question. Write the correct word or phrase on the lines provided.

1. Rewrite the question in your own words.

   _____

   _____

2. What key words do you see in the question?

   _____

3. A _____ involves moving a figure so that it is in a different position but has the same size and shape.

4. Two figures are_____ if their side and angle measures are the same.

5. Can you use geometry software to draw figures? _____

6. What transformations can you perform with geometry software?

   _____

7. Do transformations change the side or angle measures of a figure? _____

8. If a figure is transformed, is the new figure congruent to the original figure?

   _____

9. How can you prove congruence of two figures?

   _____

HOW can technology help you show the relationship between transformations and congruence?

   _____

   _____

   _____

Problem-Solving Investigation
Draw a Diagram

Case 3 Dance

Ms. Samson's dance class is evenly shaped in a **circle**.

If the **sixth person is directly opposite the sixteenth person, how many people are in the circle?**

- Understand:

- Plan:

- Solve:

- Check:

Case 4 Stadium Seating

A section of a baseball stadium is set up so that **each row has the same number of seats.**

Kyleigh is **seated in the seventh row from the back and the eighth row from the front of this section.**

Her seat is the fourth row from the right and the seventh from the left.

How many seats are in this section of the stadium?

- Understand:

- Plan:

- Solve:

- Check:

Inquiry Lab Guided Writing

Similar Triangles

HOW are two triangles related if they have the same shape but different sizes?

Use the exercises below to help answer the Inquiry Question. Write the correct word or phrase on the lines provided.

1. Rewrite the question in your own words.

_____

_____

2. What key words do you see in the question?

_____

3. Figures are similar if they have the same _____ but different

_____ .

Write *congruent* or *similar* to describe each pair of triangles.

4.  _____

5. 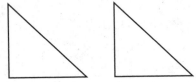 _____

6. 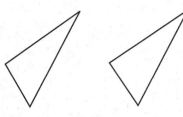 _____

7.  _____

HOW are two triangles related if they have the same shape but different sizes?

_____

_____

_____

Lesson 3 Vocabulary

Similarity and Transformations

Use the definition map to list qualities about the vocabulary word or phrase.

Vocabulary

similar

Description

**Characteristics
of the scale factor**

Draw three pairs of similar figures.

Lesson 4 Vocabulary

Properties of Similar Polygons

Use the word cards to define each vocabulary word or phrase and give an example.

Word Cards

similar polygons

Definition

_____

_____

_____

Example Sentence

_____

_____

polígonos semejantes

Definición

_____

_____

_____

Word Cards

scale factor

Definition

_____

_____

_____

Example Sentence

_____

_____

factor de escala

Definición

_____

_____

_____

Lesson 5 Vocabulary

Similar Triangles and Indirect Measurement

Use the flow chart to solve a problem using indirect measurement.

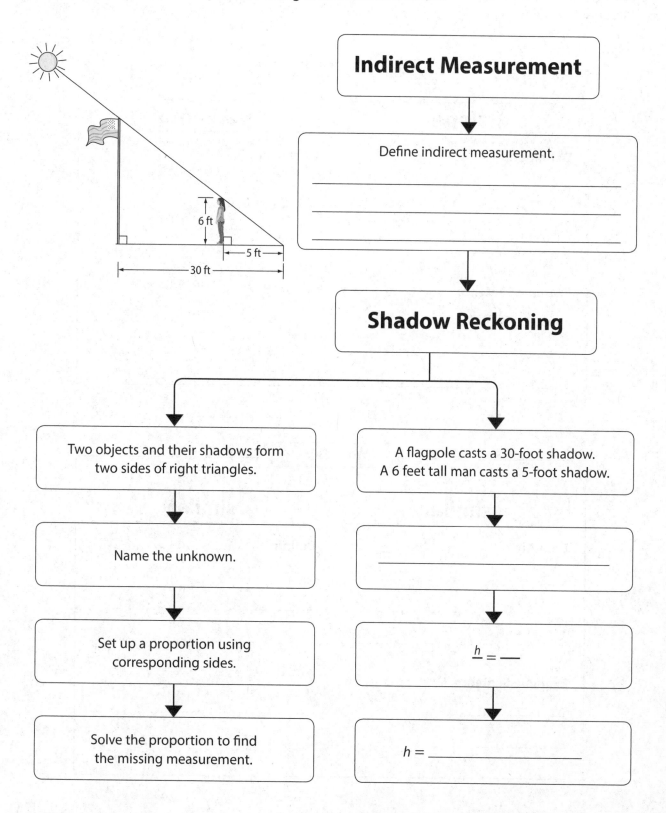

Indirect Measurement

Define indirect measurement.

_____

_____

_____

Shadow Reckoning

Two objects and their shadows form two sides of right triangles.

Name the unknown.

Set up a proportion using corresponding sides.

Solve the proportion to find the missing measurement.

A flagpole casts a 30-foot shadow.
A 6 feet tall man casts a 5-foot shadow.

_____

$$\frac{h}{\ } = \frac{\ }{\ }$$

$h = $ _____

Lesson 6 Review Vocabulary
Slope and Similar Triangles

Use the word cards to define each vocabulary word or phrase and give an example.

Word Cards

slope	**pediente**
Definition	**Definición**
_____	_____
_____	_____
_____	_____
Example Sentence	
_____	
_____	

- -

Word Cards

similar	**similar**
Definition	**Definición**
_____	_____
_____	_____
_____	_____
Example Sentence	
_____	
_____	

Lesson 7 Notetaking

Area and Perimeter of Similar Figures

Use Cornell notes to better understand the lesson's concepts. Complete each sentence by filling in the blanks with the correct word or phrase.

Questions	Notes
1. How can I use the scale factor to find the perimeter of similar figures?	If figure B is _____ to figure A by a _____ , then the perimeter of B is _____ to the perimeter of A times the _____ .
2. How can I use the scale factor to find the area of similar figures?	If figure B is _____ to figure A by a _____ , then the area of B is _____ to the area of A times the _____ of the _____ .

Summary

If you know two figures are similar and you are given the area of both figures, how can you determine the scale factor of the similarity?

_____

_____

_____

_____

_____

_____

_____

Inquiry Lab Guided Writing

Three-Dimensional Figures

HOW are some three-dimensional figures related to circles?

Use the exercises below to help answer the Inquiry Question. Write the correct word or phrase on the lines provided.

1. Rewrite the question in your own words.

   _____

   _____

2. What key words do you see in the question?

   _____

3. What words describe a circle? _____

4. Name two real-life objects that are circles. _____

5. A _____ is a figure that has length, width, and height.

6. Name three real-life three-dimensional objects that have circles as part of them.

   _____

7. What kinds of figures are those objects? _____

8. Is a circle a three-dimensional figure? _____

9. Are cylinders and spheres three-dimensional figures? _____

   HOW are some three-dimensional figures related to circles?

   _____

   _____

   _____

Lesson 1 Vocabulary
Volume of Cylinders

Use the vocabulary squares to write a definition, a sentence, and an example for each vocabulary word.

volume	Definition
Example	**Sentence**

cylinder	Definition
Draw a cylinder.	**Sentence**

composite solids	Definition
Draw a composite solid.	**Sentence**

Lesson 2 Vocabulary

Volume of Cones

Use the word cards to define each vocabulary word or phrase and give an example.

Word Cards

cone	**cono**
Definition	**Definición**
_____	_____
_____	_____
_____	_____
Example Sentence	
_____	
_____	

- -

Word Cards

vertex	**vértice**
Definition	**Definición**
_____	_____
_____	_____
_____	_____
Example Sentence	
_____	
_____	

Lesson 3 Vocabulary

Volume of Spheres

Use the definition map to list qualities about the vocabulary word or phrase.

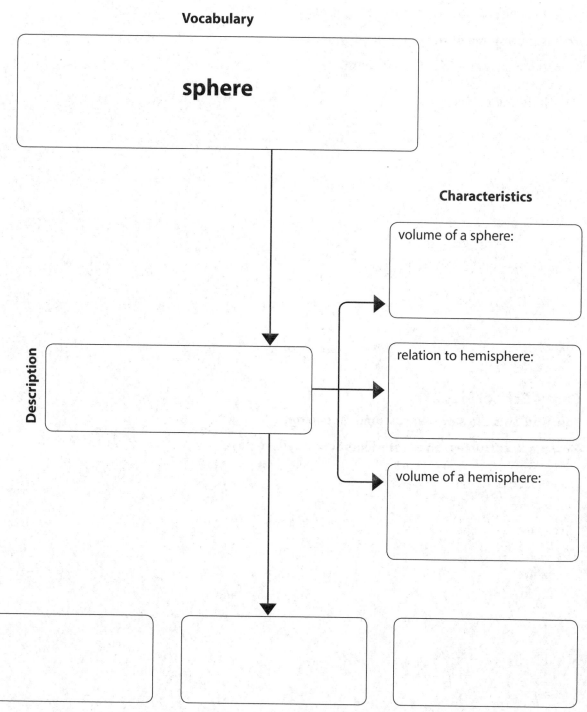

Vocabulary

sphere

Description

Characteristics

volume of a sphere:

relation to hemisphere:

volume of a hemisphere:

Draw and label three examples of spheres.

Problem-Solving Investigation

Solve a Simpler Problem

Case 3 Storage

A **15-foot tall** storage building is shown.

Grain fills the storage building to a **height of 12 feet.**

What is the **volume of the space filled** with grain?

Round your answer to the **nearest tenth.**

- Understand:

- Plan:

- Solve:

- Check:

Case 4 School Play

Four students can **sew four costumes** in **two days.**

How many costumes can **ten students sew** in **twelve days?**

- Understand:

- Plan:

- Solve:

- Check:

Inquiry Lab Guided Writing

Surface Area of Cylinders

HOW can the surface area of a cylinder be determined?

Use the exercises below to help answer the Inquiry Question. Write the correct word or phrase on the lines provided.

1. Rewrite the question in your own words.

   _____

   _____

2. What key words do you see in the question?

   _____

3. What shape is the base of a cylinder? _____

4. What is the formula for finding the area of a circle? _____

5. How many bases does a cylinder have? _____

6. What shape is the curved side of a cylinder when it is flattened? _____

7. What is the formula for finding the area of a rectangle? _____

8. To find the surface area of a cylinder, add the areas of the two _____

   and the area of the _____ .

HOW can the surface area of a cylinder be determined?

_____

_____

_____

Lesson 4 Vocabulary

Surface Area of Cylinders

Use the word cards to define each vocabulary word or phrase and give an example.

Word Cards

lateral area	área lateral
Definition	**Definición**
Example Sentence	

- -

Word Cards

total surface area	área de superficie total
Definition	**Definición**
Example Sentence	

Inquiry Lab Guided Writing
Nets of Cones

HOW can the surface area of a cone be found?

Use the exercises below to help answer the Inquiry Question. Write the correct word or phrase on the lines provided.

1. Rewrite the question in your own words.

   _____

   _____

2. What key words do you see in the question?

   _____

3. How many bases does a cone have? _____

4. What shape is the base of a cone? _____

5. What is the formula for finding the area of a circle? _____

6. The area that forms the curved side of a cone is the

   _____ surface area.

7. The formula $A = \pi r \ell$ is used for finding the lateral

   surface area of a _____ .

8. What does the "ℓ" stand for in the formula $A = \pi r \ell$? _____

9. A cone has a base surface area of 2.46 cm² and a lateral surface area of 18.84 cm².

   How do you find the total surface area of the cone? _____ .

HOW can the surface area of a cone be found?

_____

_____

_____

Lesson 5 Notetaking

Surface Area of Cones

Use Cornell notes to better understand the lesson's concepts. Complete each sentence by filling in the blanks with the correct word or phrase.

Questions	Notes
1. How do I find the lateral area of a cone?	The lateral area of a cone is _____ times the _____ times ℓ, the _____ .
2. How do I find the surface area of a cone?	The surface area of a cone is the _____ plus the _____ of the circular base.

Summary

How does the volume of a three-dimensional figure differ from its surface area?

_____

_____

_____

_____

_____

_____

Inquiry Lab Guided Writing

Changes in Scale

HOW does multiplying the dimensions of a three-dimensional figure by a scale factor affect its volume and surface area?

Use the exercises below to help answer the Inquiry Question. Write the correct word or phrase on the lines provided.

1. Rewrite the question in your own words.

_____

_____

2. What key words do you see in the question?

_____

3. A _____ is the ratio of an enlarged or reduced figure to the original figure.

4. _____ is measured in cubic units.

_____ is measured in square units.

Use the table to complete Exercises 5-7.

	Small Cube	Large Cube
Volume	125 in³	1,000 in³
SurfaceArea	150 in²	600 in²

Scale factor = 2

5. The volume of the small cube is multiplied by _____ to find the volume of the large cube.

6. The surface area of the small cube is multiplied by _____ to find the surface area of the large cube.

7. What is the scale factor cubed? _____

What is the scale factor squared? _____

HOW does multiplying the dimensions of a three-dimensional figure by a scale factor affect its volume and surface area?

_____

_____

_____

Lesson 6 Vocabulary

Changes in Dimensions

Use the word cards to define each vocabulary word or phrase and give an example.

Word Cards

| **similar solids** | **sólidos semejantes** |

Definition

_____

_____

_____

Definición

_____

_____

_____

Example Sentence

_____

_____

Word Cards

| **scale factor** | **factor de escala** |

Definition

_____

_____

_____

Definición

_____

_____

_____

Example Sentence

_____

_____

Inquiry Lab Guided Writing

Scatter Plots

HOW can I use a graph to investigate the relationship or trends between two sets of data?

Use the exercises below to help answer the Inquiry Question. Write the correct word or phrase on the lines provided.

1. Rewrite the question in your own words.

_____

_____

2. What key words do you see in the question?

_____

3. A pair of numbers used to locate a point in the coordinate plane is called an

_____ .

4. Write a synonym for the word *trend*. _____

5. Write the following data in the table as ordered pairs. Then graph the ordered pairs on the coordinate plane.

Distance Run (miles)	Calories Burned	Ordered Pair
0.5	49	
1	98	
1.5	147	
2	196	
2.5	245	

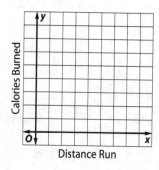

Distance Run

6. Does the graph show a trend in the data? _____

If yes, describe the trend. _____

HOW can I use a graph to investigate the relationship or trends between two sets of data?

_____

_____

_____

Lesson 1 Vocabulary

Scatter Plots

Use the word cards to define each vocabulary word or phrase and give an example.

Word Cards

bivariate data	datos bivariantes
Definition	**Definición**
_____	_____
_____	_____
_____	_____
Example Sentence	
_____	
_____	

- -

Word Cards

scatter plot	diagram de dispersión
Definition	**Definición**
_____	_____
_____	_____
_____	_____
Example Sentence	
_____	
_____	

Inquiry Lab Guided Writing

Lines of Best Fit

HOW can I use a data model to predict an outcome?

Use the exercises below to help answer the Inquiry Question. Write the correct word or phrase on the lines provided.

1. Rewrite the question in your own words.

_____

_____

2. What key words do you see in the question?

_____

3. _____ means to tell what you think will happen.

4. Write a synonym for the word *outcome*. _____

5. Complete the steps on how to use a data model to predict an outcome.

a. Conduct research to collect a set of _____ .

b. Write the data in the form of _____ pairs.

c. Create a _____ by marking the points in a coordinate plane.

d. Draw a _____ that goes through most of the data points.

e. Make a _____ based on the line you drew.

HOW can I use a data model to predict an outcome?

_____

_____

_____

Lesson 2 Vocabulary
Lines of Best Fit

Use the definition map to list qualities about the vocabulary word or phrase.

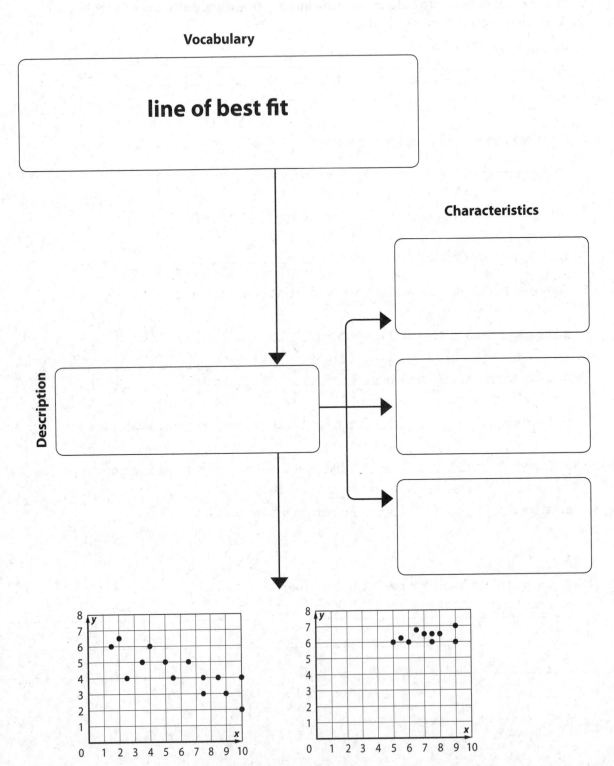

Vocabulary

line of best fit

Characteristics

Description

Draw the lines of best fit.

Inquiry Lab Guided Writing

Graphing Technology: Linear and Nonlinear Association

HOW can you use technology to describe associations in scatter plots?

Use the exercises below to help answer the Inquiry Question. Write the correct word or phrase on the lines provided.

1. Rewrite the question in your own words.

   _____

   _____

2. What key words do you see in the question?

   _____

3. A _____ shows two sets of related data as ordered pairs on the same graph.

4. A _____ is an electronic tool you can use to create a scatter plot of data.

5. A line that is very close to most of the data points is called a _____ .

6. Write a synonym for the word *associations*. _____

7. The graph of a linear association is a _____ line.

8. The correlation coefficient tells the _____ of the association between two sets of data.

9. If data is clustered closely around the line of best fit, the strength of the association is _____ .

10. If the data is not clustered closely around the line of best fit, the association is

    _____ .

HOW can you use technology to describe associations in scatter plots?

_____

_____

_____

Lesson 3 Vocabulary

Two-Way Tables

Use the word cards to define each vocabulary word or phrase and give an example.

Word Cards

relative frequency	frecuencia relativa
Definition	**Definición**
_____	_____
_____	_____
_____	_____
Example Sentence	
_____	
_____	

Word Cards

two-way table	table de doble entrada
Definition	**Definición**
_____	_____
_____	_____
_____	_____
Example Sentence	
_____	
_____	

Problem-Solving Investigation
Use a Graph

Case 3 Blogs

The numbers of followers of a popular blog are shown in the table.

What is a reasonable **estimate** for the **number of followers in Year 10** if this trend continues?

Year	Number of Followers
1	42,000
2	50,000
3	76,000
4	94,000
5	115,000

- Understand:

- Plan:

- Solve:

- Check:

Case 4 School Colors

The graph shows the results of a favorite color survey.

To the nearest percent, what **percent more** of the students **chose purple and orange than green and gray**?

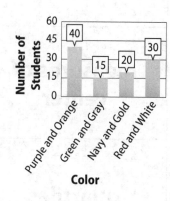

- Understand:

- Plan:

- Solve:

- Check:

Lesson 4 Vocabulary

Descriptive Statistics

Use the three column chart to organize the vocabulary in this lesson. Write the word in Spanish. Then write the definition of each word.

English	Spanish	Definition
univariate data		
quantitative data		
five-number summary		
measures of center		
quartiles		

Lesson 5 Vocabulary
Measures of Variation

Use the word cards to define each vocabulary word or phrase and give an example.

Word Cards

mean absolute deviation

Definition

_____

_____

_____

Example Sentence

_____

_____

desviación media absoluta

Definición

_____

_____

_____

Word Cards

standard deviation

Definition

_____

_____

_____

Example Sentence

_____

_____

desviación estándar

Definición

_____

_____

_____

Lesson 6 Vocabulary
Analyze Data Distributions

Use the definition map to list qualities about the vocabulary word or phrase.

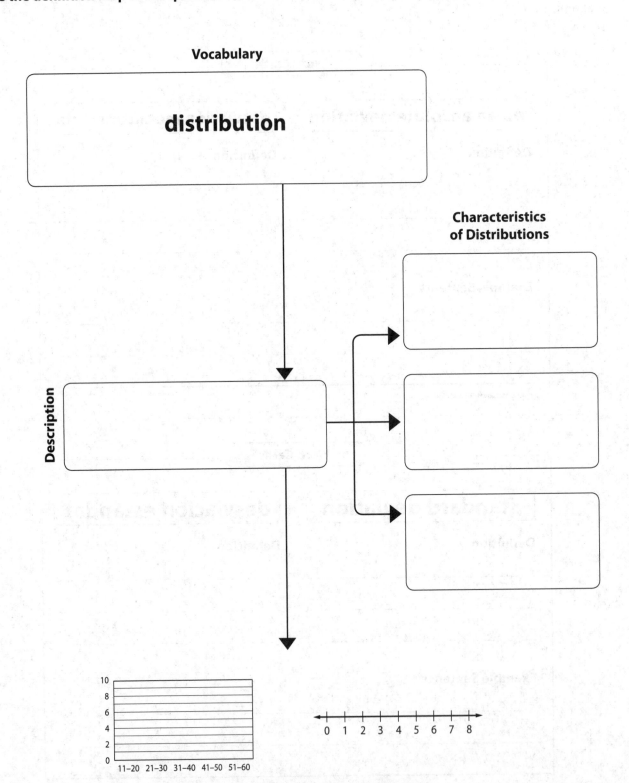

Vocabulary

distribution

**Characteristics
of Distributions**

Description

Draw Examples of Symmetric Distributions

What are VKVs® and How Do I Create Them?

Visual Kinethestic Vocabulary Cards® are flashcards that animate words by focusing on their structure, use, and meaning. The VKVs in this book are used to show cognates, or words that are similar in Spanish and English.

Step 1

Go to the back of your book to find the VKVs for the chapter vocabulary you are currently studying. Follow the cutting and folding instructions at the top of the page. The vocabulary word on the BLUE background is written in English. The Spanish word is on the ORANGE background.

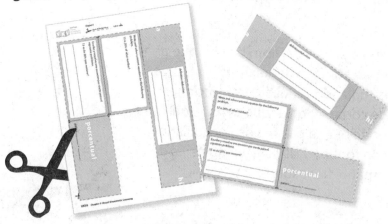

Step 2

There are exercises for you to complete on the VKVs. When you understand the concept, you can complete each exercise. All exercises are written in English and Spanish. You only need to give the answer once.

Step 3

Individualize your VKV by writing notes, sketching diagrams, recording examples, and forming plurals (radius: radii or radiuses).

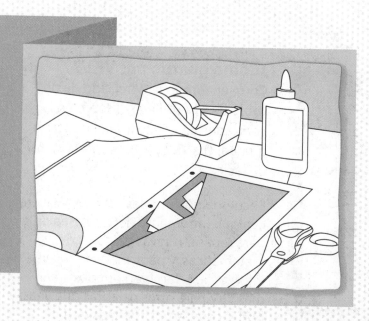

How Do I Store My VKVs?

Take a 6" x 9" envelope and cut away a V on one side only. Glue the envelope into the back cover of your book. Your VKVs can be stored in this pocket!

Remember you can use your VKVs ANY time in the school year to review new words in math, and add new information you learn. Why not create your own VKVs for other words you see and share them with others!

Las tarjetas de vocabulario visual y cinético (VKV) contienen palabras con animación que está basada en la estructura, uso y significado de las palabras. Las tarjetas de este libro sirven para mostrar cognados, que son palabras similares en español y en inglés.

Paso 1

Busca al final del libro las VKV que tienen el vocabulario del capítulo que estás estudiando. Sigue las instrucciones de cortar y doblar que se muestran al principio. La palabra de vocabulario con fondo AZUL está en inglés. La de español tiene fondo NARANJA.

Paso 2

Hay ejercicios para que completes con las VKV. Cuando entiendas el concepto, puedes completar cada ejercicio. Todos los ejercicios están escritos en inglés y español. Solo tienes que dar la respuesta una vez.

Paso 3

Da tu toque personal a las VKV escribiendo notas, haciendo diagramas, grabando ejemplos y formando plurales (radio: radios).

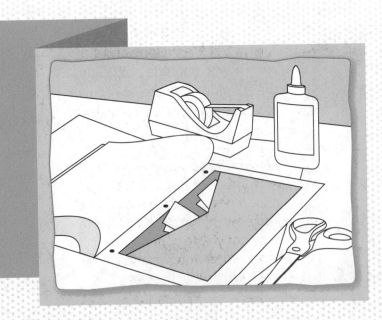

¿Cómo guardo mis VKV?

Corta en forma de "V" el lado de un sobre de 6" X 9". Pega el sobre en la contraportada de tu libro. Puedes guardar tus VKV en esos bolsillos. ¡Así de fácil!

Recuerda que puedes usar tus VKV en cualquier momento del año escolar para repasar nuevas palabras de matemáticas, y para añadir la nueva información. También puedes crear más VKV para otras palabras que veas, y poder compartirlas con los demás.

✂ cut on all dashed lines

▭ fold on all solid lines

Define exponent. (Define exponente.)

_____

_____

_____

exponente

b^x

base

Define base. (Define base.)

_____

_____

_____

racional

Circle the irrational number.

Spanish Translation

$\frac{5}{8}$ −6.5   12%

$\sqrt{8}$ 0.2222...

irrational number

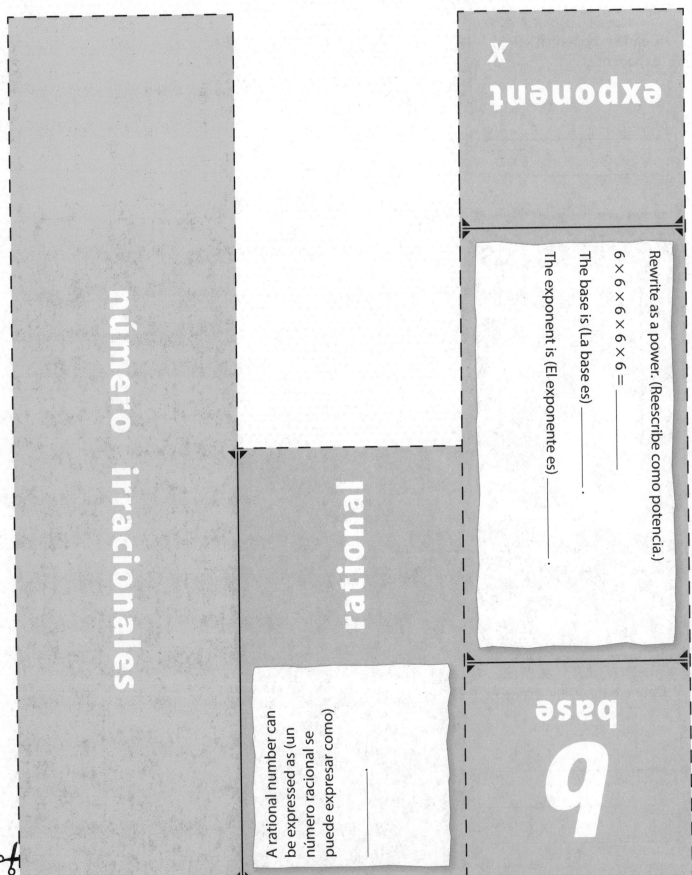

número irracionales

rational

A rational number can be expressed as (un número racional se puede expresar como)

_____

exponent

x

Rewrite as a power. (Reescribe como potencia.)

$6 \times 6 \times 6 \times 6 \times 6 =$ _____

The base is (La base es) _____ .

The exponent is (El exponente es) _____ .

base

6

Write about a time when you would use scientific notation. (Escribe sobre una situación en la cual usarías la notación científica.)

_____

Circle the perfect cubes. (Encierra en un círculo los cubos perfectos.)

6 9 27

125 200 625

Circle the radical sign. (Encierra en un círculo el signo radical.)

$\sqrt{}$ +

×

scientific notation

perfect cube

radical sign

Circle the numbers written in scientific notation. (Encierra en un círculo los números escritos en notación científica.)

0.034 2.75×10^5

3×10^{-4} 98.3

Find each cube root. (Halla la raíz cúbica.)

$\sqrt[3]{-27} =$ _____

$\sqrt[3]{216} =$ _____

$\sqrt[3]{8,000} =$ _____

Simplify the expression. (Simplifica la expresión.)

$\sqrt[3]{216} =$ _____

Dinah Zike's
Visual
Kinesthetic
Vocabulary®

Chapter 1

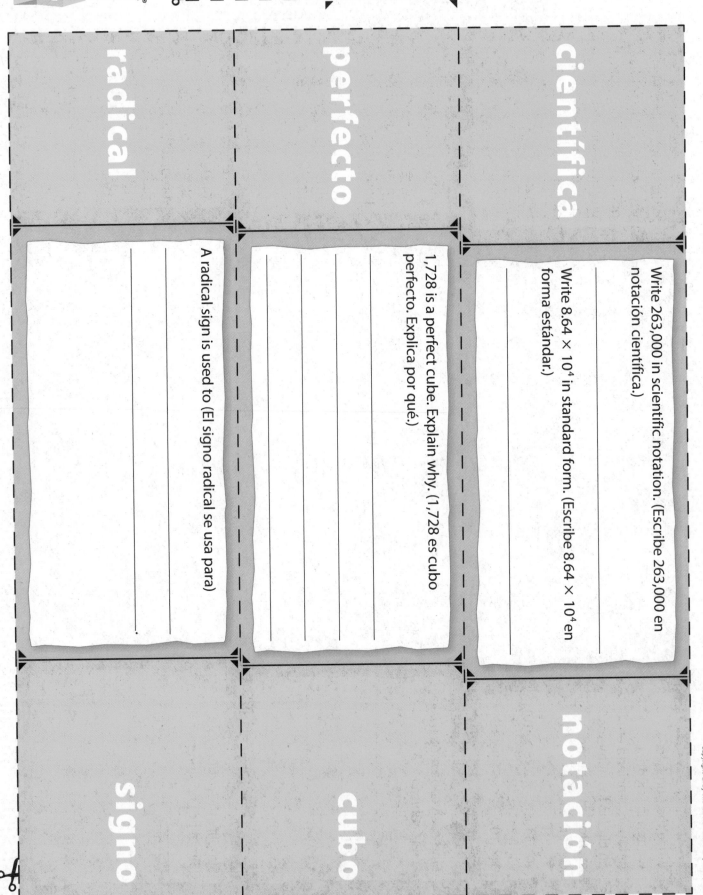

✂ cut on all dashed lines   ⬚ fold on all solid lines

radical

perfecto

científica

A radical sign is used to (El signo radical se usa para)

1,728 is a perfect cube. Explain why. (1,728 es cubo perfecto. Explica por qué.)

Write 263,000 in scientific notation. (Escribe 263,000 en notación científica.)

Write 8.64×10^4 in standard form. (Escribe 8.64×10^4 en forma estándar.)

signo

cubo

notación

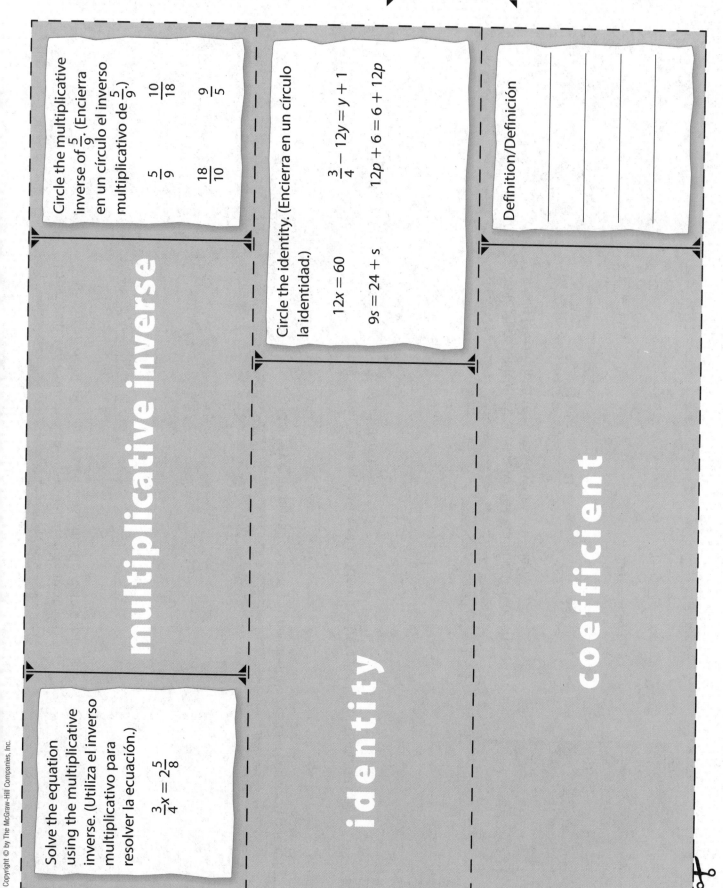

Circle the multiplicative inverse of $\frac{5}{9}$. (Encierra en un círculo el inverso multiplicativo de $\frac{5}{9}$.)

$\frac{10}{18}$ $\frac{9}{5}$

$\frac{5}{9}$ $\frac{18}{10}$

Circle the identity. (Encierra en un círculo la identidad.)

$\frac{3}{4} - 12y = y + 1$

$12x = 60$ $12p + 6 = 6 + 12p$

$9s = 24 + s$

Definition/Definición

multiplicative inverse

Solve the equation using the multiplicative inverse. (Utiliza el inverso multiplicativo para resolver la ecuación.)

$\frac{3}{4}x = 2\frac{5}{8}$

identity

coefficient

Dinah Zike's
Visual
Kinesthetic
Vocabulary®

Chapter 2

✂ cut on all dashed lines

fold on all solid lines

multiplicativo

dad

e

The product of a number and its multiplicative inverse is (El producto de un número y su inverso multiplicativo es)

_____ .

An identity is an equation that is (Una identidad es una ecuación que es)

_____ .

What is the opposite of an identity? (¿Que es lo contrario de una identidad?)

_____ .

Circle the coefficients. Then solve the equations. (Encierra en un círculo los coeficientes y resuelve las ecuacións.)

$6x = 19.2$

$x =$ _____

$3\frac{1}{2} = 14m$

$m =$ _____

$25 = \left(\frac{5}{8}\right)y$

$y =$ _____

$5.4p = 48.6$

$p =$ _____

inverso

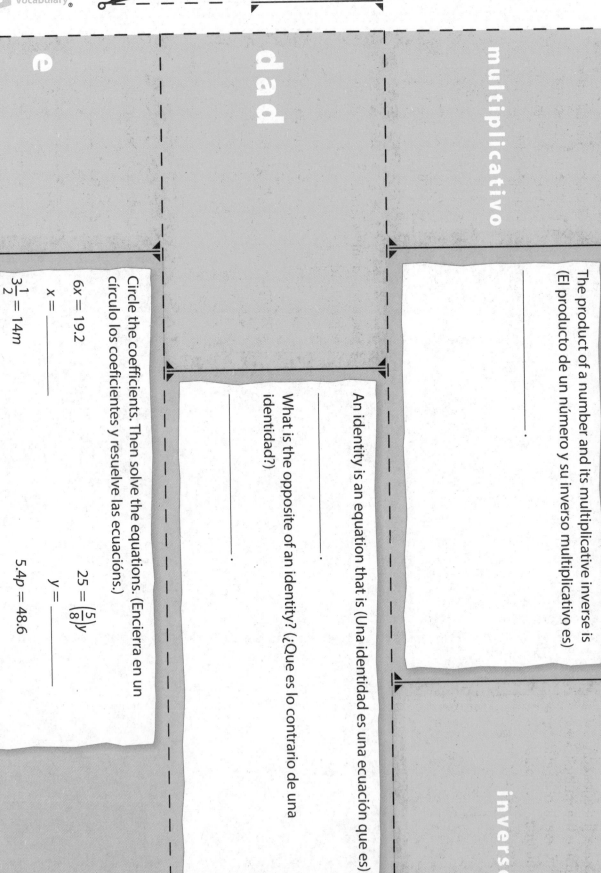

Dinah Zike's Visual Kinesthetic Vocabulary®

cut on all dashed lines

fold on all solid lines

Why might it be useful to write an equation in standard form? (¿Qué utilidad tendría escribir una ecuación de forma estándar?)

Define linear relationship. (Define relación lineal.)

standard form

linear relationship

What does a linear relationship look like when it is graphed? (¿Cómo se ve la gráfica de una relación lineal?)

Dinah Zike's
VKV Visual
Kinesthetic
Vocabulary®

Chapter 3

✂ cut on all dashed lines

⬜ fold on all solid lines

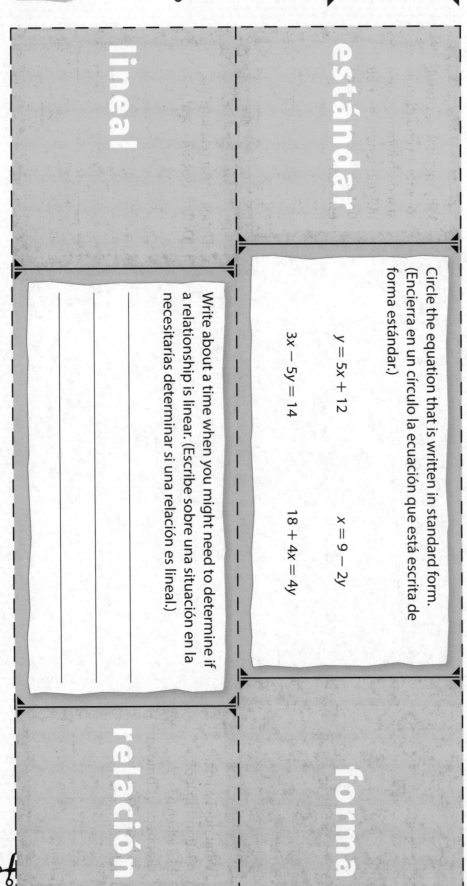

lineal

estándar

Circle the equation that is written in standard form. (Encierra en un círculo la ecuación que está escrita de forma estándar.)

$y = 5x + 12$

$x = 9 - 2y$

$3x - 5y = 14$

$18 + 4x = 4y$

Write about a time when you might need to determine if a relationship is linear. (Escribe sobre una situación en la necesitarías determinar si una relación es lineal.)

relación

forma

Define substitution. (Define sustitución.)

substitution

system of equations

sistema de ecuaciones

stitución

Solve the system of equations. (Resuelve el sistema de ecuaciones.)

$y = 2x - 7$
$y = 5 - 4x$

(,)

Use substitution to solve the system of equations. (Resuelve el sistema de ecuaciones mediante el método de sustitución.)

$y = 3x$
$y = 5x - 6$

(,)

Define system of equations. (Define sistema de ecuaciones.)

Dinah Zike's
Visual
Kinesthetic
Vocabulary®

Chapter 4

✂ ----- cut on all dashed lines

▸▭ fold on all solid lines ◂

Define relation. (Define relación.)

Define range. (Define rango.)

Define domain. (Define dominio.)

relation

range

domain

inio

o

ción

Find four ordered pairs for the relation $y = x + 4$. (Enumera cuatro pares ordenados de la relación $y = x + 4$.)

(_____ , _____),   (_____ , _____),

(_____ , _____),   (_____ , _____)

State the range of the relation below. (Calcula el rango de la siguiente relación.)

$\{(6, -2), (8, 0), (12, 2), (18, 6)\}$

{ _____ , _____ , _____ , _____ }

State the domain of the relation below. (Calcula el dominio de la siguiente relación.)

$\{(-5, 4), (-7, 3), (-12, 11), (-14, 13)\}$

{ _____ , _____ , _____ , _____ }

Dinah Zike's
Visual
Kinesthetic
Vocabulary ®

Chapter 4

✂ cut on all dashed lines   ⬜ fold on all solid lines

Draw an example of a quadratic function below. (Dibuja un ejemplo de una función cuadrática.)

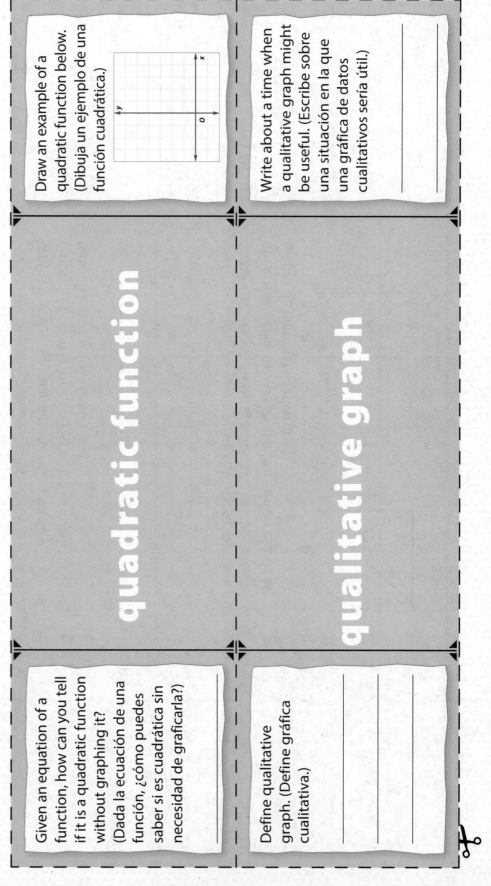

Write about a time when a qualitative graph might be useful. (Escribe sobre una situación en la que una gráfica de datos cualitativos sería útil.)

quadratic function

qualitative graph

Given an equation of a function, how can you tell if it is a quadratic function without graphing it? (Dada la ecuación de una función, ¿cómo puedes saber si es cuadrática sin necesidad de graficarla?)

Define qualitative graph. (Define gráfica cualitativa.)

cualitativa

cuadrática

Why is $y = x^2 + 5$ a quadratic function, while $y = x^3 - 9$ is not? (¿Por qué la función $y = x^2 + 5$ es cuadrática y la función $y = x^3 - 9$ no lo es?)

Circle the qualitative graph. (Encierra en un círculo la gráfica de datos cualitativos.)

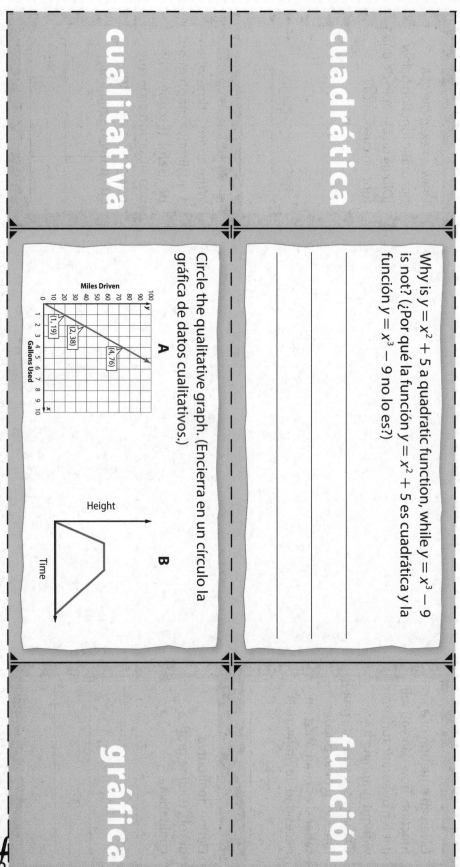

A

B

gráfica

función

cut on all dashed lines

fold on all solid lines

nonlinear function

lineal

Draw an example of a nonlinear function on the graph at the right. (Dibuja un ejemplo de una función no lineal en la gráfica de la derecha.)

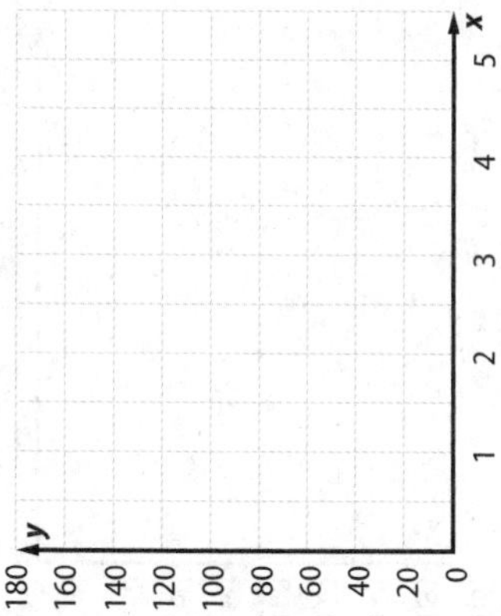

y

180
160
140
120
100
80
60
40
20

0   1   2   3   4   5   x

Dinah Zike's
Visual
Kinesthetic
Vocabulary ®

Chapter 4

✂ cut on all dashed lines

fold on all solid lines

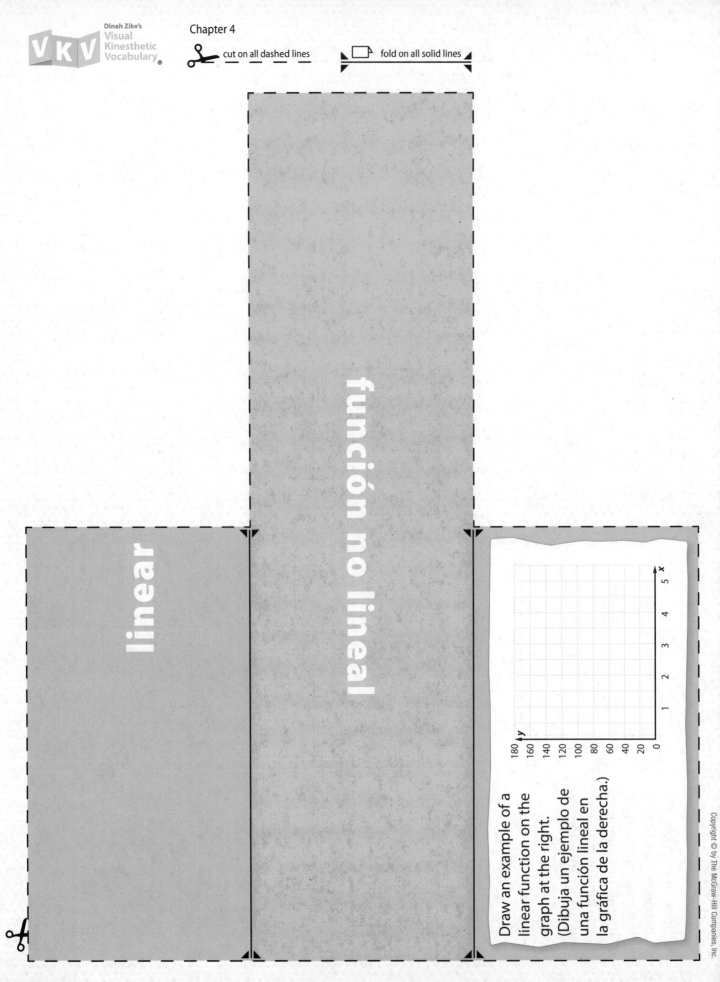

linear

función no lineal

Draw an example of a
linear function on the
graph at the right.
(Dibuja un ejemplo de
una función lineal en
la gráfica de la derecha.)

✂ cut on all dashed lines   fold on all solid lines

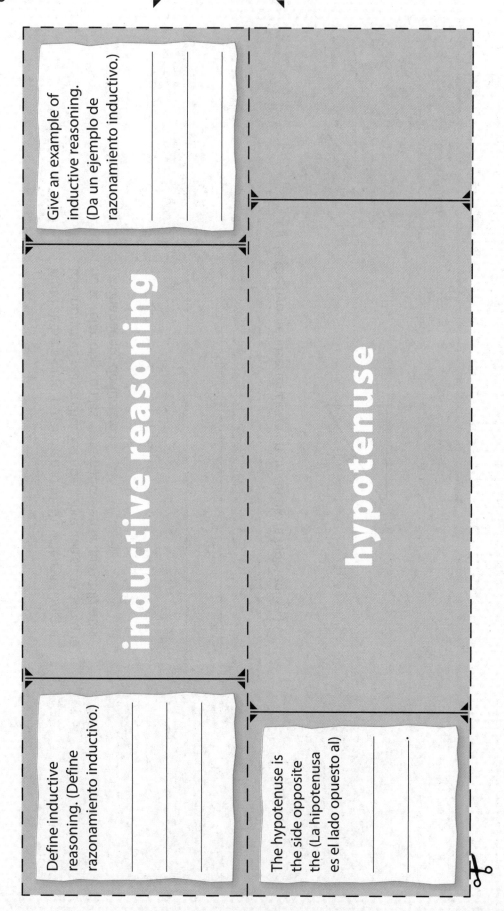

Give an example of inductive reasoning. (Da un ejemplo de razonamiento inductivo.)

inductive reasoning

hypotenuse

Define inductive reasoning. (Define razonamiento inductivo.)

The hypotenuse is the side opposite the (La hipotenusa es el lado opuesto al)

a

inductivo

razonamiento

Circle the hypotenuse. (Encierra en un círculo la hipotenusa.)

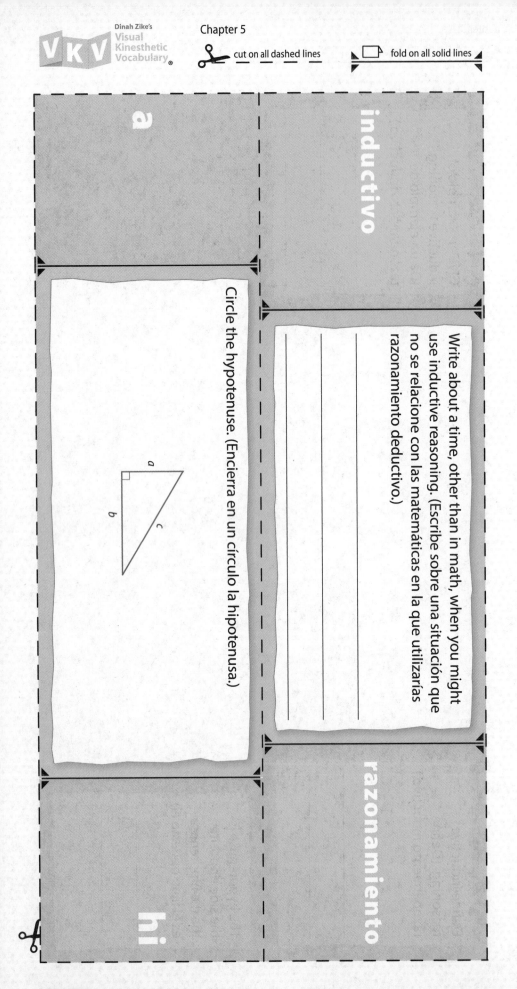

Write about a time, other than in math, when you might use inductive reasoning. (Escribe sobre una situación que no se relacione con las matemáticas en la que utilizarías razonamiento deductivo.)

hi.

✂ cut on all dashed lines

⬚ fold on all solid lines

Pythagorean Theorem

Write the Pythagorean Theorem. (Escribe el teorema de Pitágoras.)

$$\underline{\hspace{2cm}} + \underline{\hspace{2cm}} = \underline{\hspace{2cm}}$$

✂ cut on all dashed lines

⬚ fold on all solid lines

Dinah Zike's
Visual
Kinesthetic
Vocabulary ®

✂ cut on all dashed lines

fold on all solid lines

Teorema de Pitágoras

Use the Pythagorean Theorem to find c. (Calcula c con ayuda del teorema de Pitágoras.)

16 m

c m

12 m

c = _____

Dinah Zike's
VKV Visual
Kinesthetic
Vocabulary ®

Chapter 5

✂ cut on all dashed lines

fold on all solid lines

Use the figure below to complete the following equation. (Completa la ecuación con ayuda de la siguiente figura.)

$$m\angle 6 = m\angle \underline{\qquad} + m\angle \underline{\qquad}$$

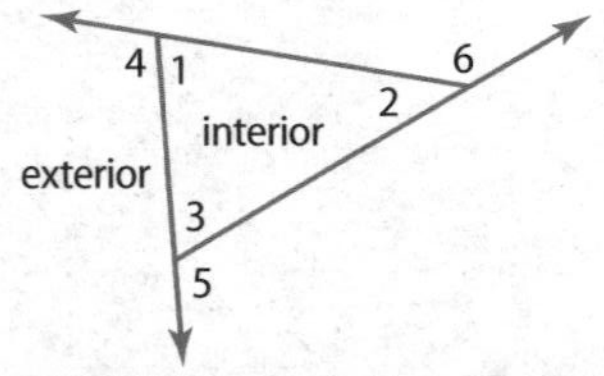

triangle

Draw an example of a regular polygon. (Dibuja un ejemplo de un polígono regular.)

regular polygon

cut on all dashed lines

fold on all solid lines

ángulo

A triangle has _______ vertices and _______ interior angles. The sum of the interior angles is _______°. (Un triángulo tiene _______ vértices y _______ ángulos internos. La suma de la medida de los ángulos internos es _______°.)

polígono regular

Define polygon. (Define polígono.)

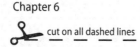

Dinah Zike's
Visual
Kinesthetic
Vocabulary®

Chapter 6

✂ cut on all dashed lines

▭ fold on all solid lines

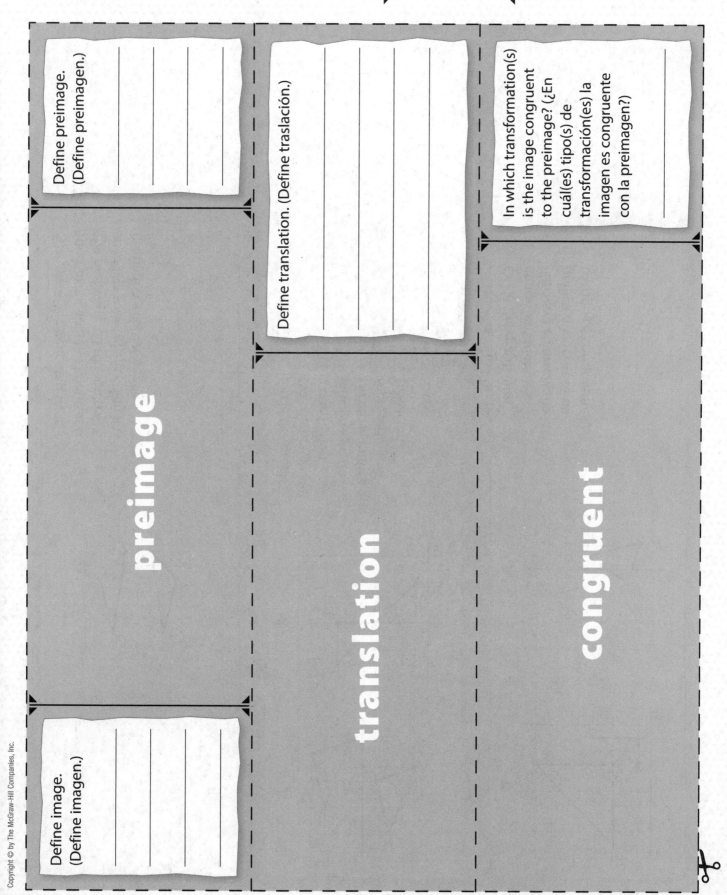

Define preimage.
(Define preimagen.)

Define translation. (Define traslación.)

In which transformation(s) is the image congruent to the preimage? (¿En cuál(es) tipo(s) de transformación(es) la imagen es congruente con la preimagen?)

preimage

translation

congruent

Define image.
(Define imagen.)

Dinah Zike's
Visual
Kinesthetic
Vocabulary®

✂ cut on all dashed lines   ▭ fold on all solid lines

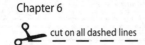

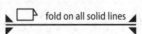
slación

n

pre

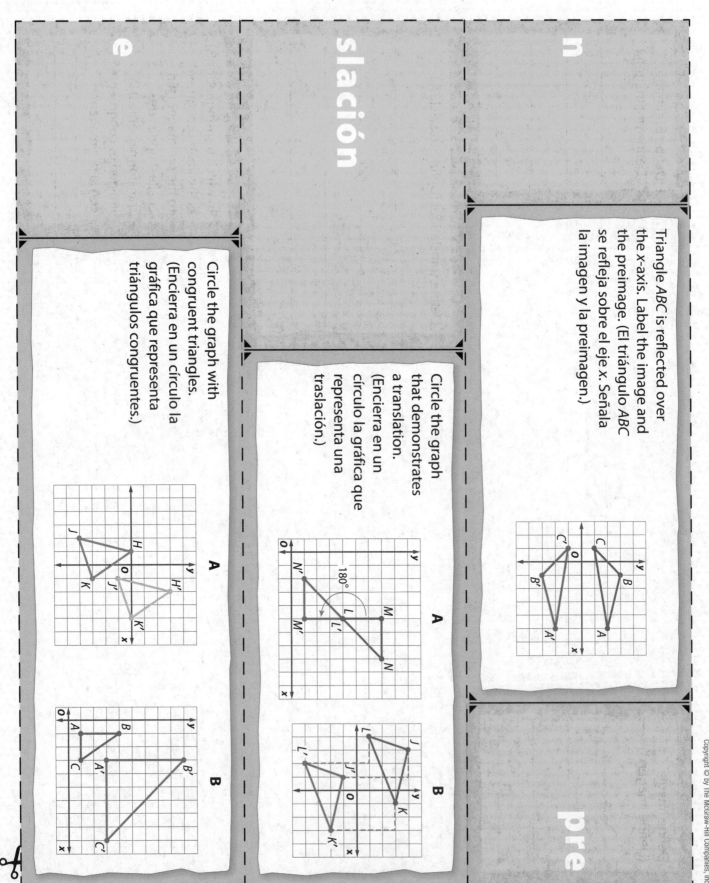

Triangle ABC is reflected over the x-axis. Label the image and the preimage. (El triángulo ABC se refleja sobre el eje x. Señala la imagen y la preimagen.)

Circle the graph that demonstrates a translation. (Encierra en un círculo la gráfica que representa una traslación.)

Circle the graph with congruent triangles. (Encierra en un círculo la gráfica que representa triángulos congruentes.)

cut on all dashed lines

fold on all solid lines

Define transformation. (Define transformación.)

transformation

Identify the line of reflection. (Identifica el línea de reflexión.)

line of reflection

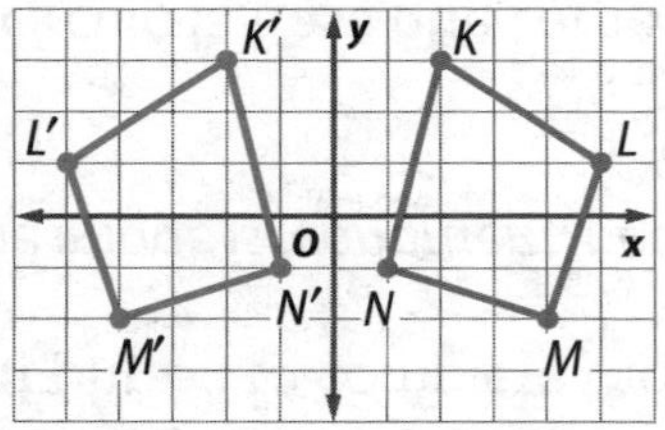

Dinah Zike's
Visual
Kinesthetic
Vocabulary ®

Chapter 6

✂ cut on all dashed lines

📄 fold on all solid lines

...ción

línea de reflexión

Describe the transformation shown in each figure. (Describe la transformación que se muestra en cada figura.)

A

B

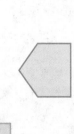

Circle the correct phrase to complete the sentence below.

In a reflection, the image is (congruent, not congruent) to the preimage.

(Encierra en un círculo la frase que completa correctamente la siguiente oración.)

En un reflexión, la imagen (es congruente, no es congruente) con la preimagen.

Dinah Zike's
**Visual
Kinesthetic
Vocabulary**®

center of rotation

ángulo de

Circle the center of rotation. (Encierra en un círculo el centro de rotación.)

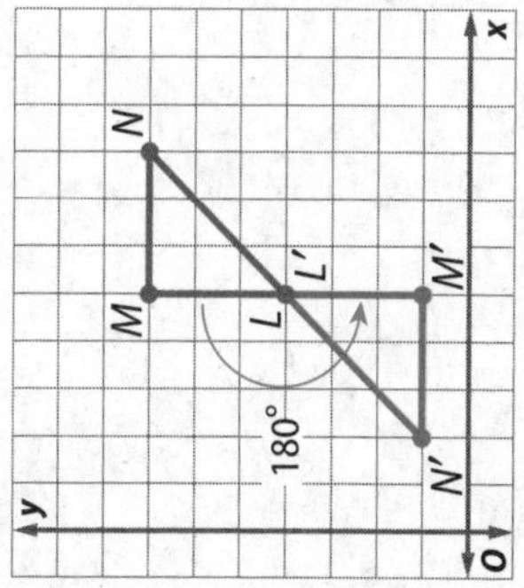

Dinah Zike's
VKV
Visual
Kinesthetic
Vocabulary ®

Chapter 6

✂ cut on all dashed lines

◻ fold on all solid lines

angle of

centro de rotación

Find the angle of rotation. (Calcula el ángulo de rotación.)

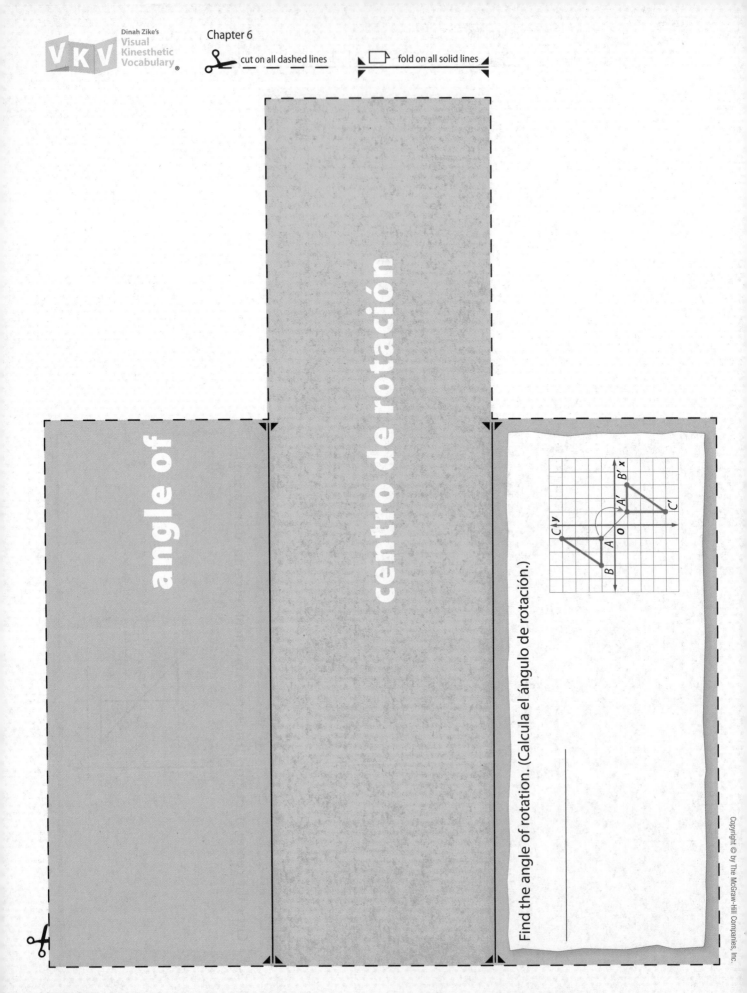

Chapter 7

Dinah Zike's
Visual
Kinesthetic
Vocabulary ®

✂ cut on all dashed lines

▭ fold on all solid lines

Encierra en un círculo la frase que completa correctamente la siguiente oración.

Si el factor de escala es menor que 1, la imagen es (más grande, más pequeña) que la preimagen.

¿Cómo puedes demostrar que en dos figuras congruentes los lados correspondientes son congruentes?

scale factor

corresponding parts

Circle the correct phrase to complete the sentence below.

If the scale factor is less than 1, the image is (an enlargement, a reduction) of the preimage.

In two congruent figures, how do you show that the corresponding sides are congruent?

cut on all dashed lines

fold on all solid lines

escala

Triangles *ABC* and *DEF* are similar. If $BC = 8$ cm, $EF = 2$ cm, and $CA = 12$ cm, what is *FD*? (Los triángulos *ABC* y *DEF* son semejantes. Si $BC = 8$ cm, $EF = 2$ cm y $CA = 12$ cm, ¿cuánto mide *FD*?)

$FD =$ ____________

factor de

correspondientes

Triangles *ABC* and *DEF* are congruent. Name two pairs of corresponding parts. (Los triángulos *ABC* y *DEF* son congruentes. Menciona dos pares de partes correspondientes.)

____________________ ____________________

partes

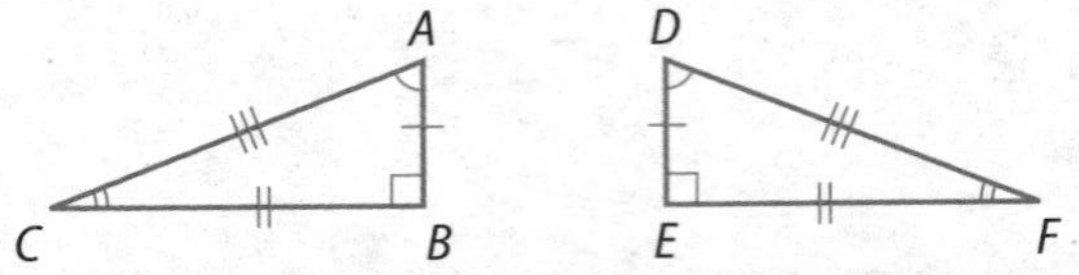

cut on all dashed lines

fold on all solid lines

Dinah Zike's
Visual
Kinesthetic
Vocabulary®

Define cylinder. (Define cilindro.)

Shade the lateral area of the cone. Then find the lateral area. (Sombrea el área lateral del cono. Luego calcula el área lateral.)

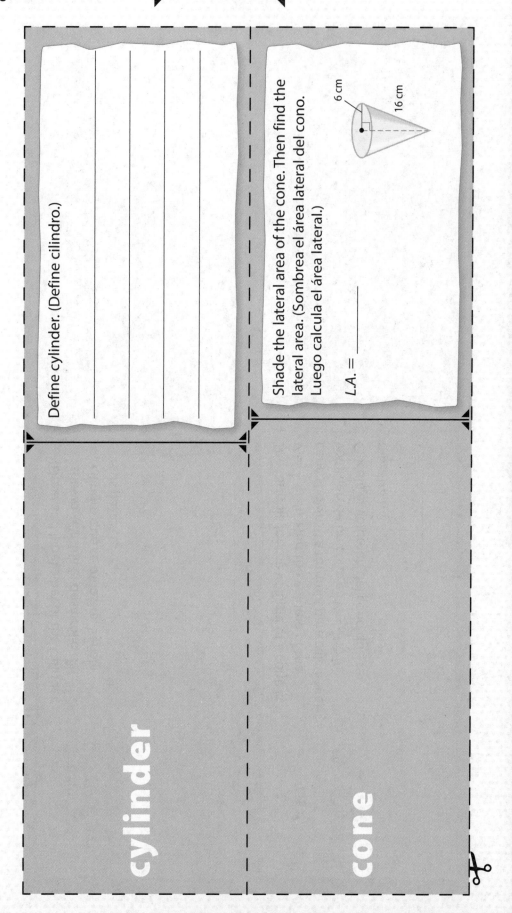

6 cm

16 cm

L.A. = _____

cylinder

cone

o

ilindro

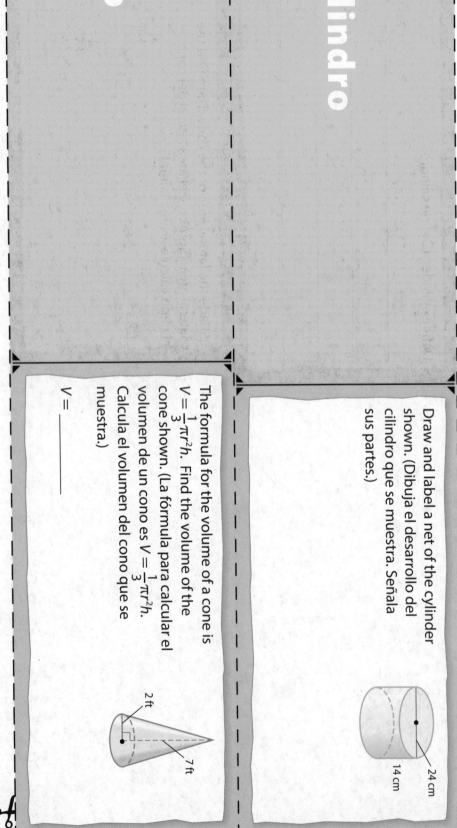

Draw and label a net of the cylinder shown. (Dibuja el desarrollo del cilindro que se muestra. Señala sus partes.)

24 cm

14 cm

The formula for the volume of a cone is $V = \frac{1}{3}\pi r^2 h$. Find the volume of the cone shown. (La fórmula para calcular el volumen de un cono es $V = \frac{1}{3}\pi r^2 h$. Calcula el volumen del cono que se muestra.)

$V = $ _____

2 ft

7 ft

Dinah Zike's
Visual
Kinesthetic
Vocabulary®

✂ cut on all dashed lines

▭ fold on all solid lines

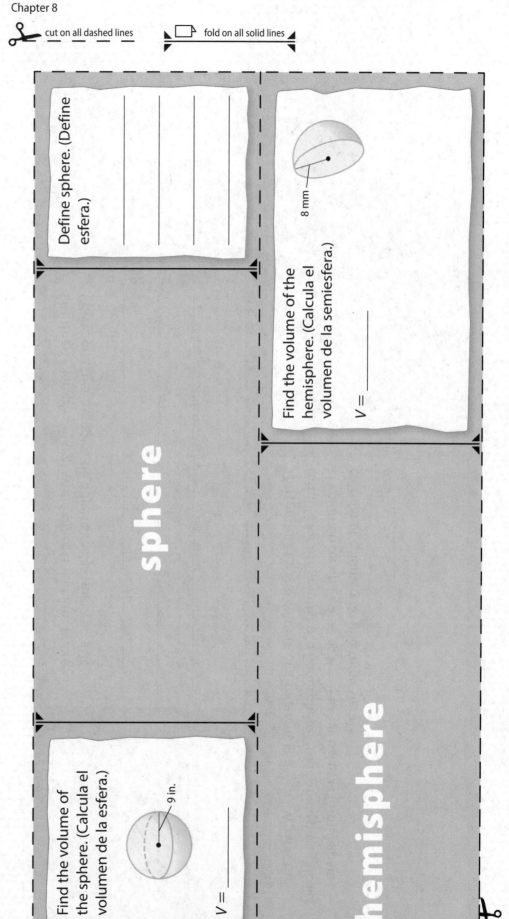

Define sphere. (Define esfera.)

Find the volume of the hemisphere. (Calcula el volumen de la semiesfera.)

8 mm

$V =$ _____

sphere

hemisphere

Find the volume of the sphere. (Calcula el volumen de la esfera.)

9 in.

$V =$ _____

Dinah Zike's
VKV Visual
Kinesthetic
Vocabulary®

Chapter 8

✂ cut on all dashed lines

⬚ fold on all solid lines

ferio

a

List three examples of real-world objects that are spheres. (Menciona tres ejemplos de objetos reales que sean esferas.)

esf

The formula for the volume of a sphere is $V = \frac{4}{3}\pi r^3$. What is the formula for the volume of a hemisphere? (La fórmula para calcular el volumen de una esfera es $V = \frac{4}{3}\pi r^3$. ¿Cuál es la fórmula para calcular el volumen de una semiesfera?)

$V = \underline{\hspace{2cm}}$

Dinah Zike's
Visual
Kinesthetic
Vocabulary ®

Chapter 9

✂ cut on all dashed lines

fold on all solid lines

Define symmetric. (Define simétrico.)

Define standard deviation. (Define desviación estándar.)

La frecuencia relativa de personas que tienen dos gatos con respecto a las personas que tienen al menos un gato es 0.83. Hay _____ personas con al menos un gato que tienen otro.

symmetric

standard deviation

relative frequency

The relative frequency of people who own two cats compared to those who have at least one cat is 0.83. _____ people with at least one cat have two.

relativa

estándar

imétrico

Define relative frequency. (Define frecuencia relativa.)

Standard deviation relates to which measure: mode, median, mean, or range? (¿Con qué medida se relaciona la desviación estándar: moda, mediana, media o rango?)

On the number line, draw an example of a distribution that is symmetric. (Dibuja un ejemplo de distribución simétrica en la recta numérica.)

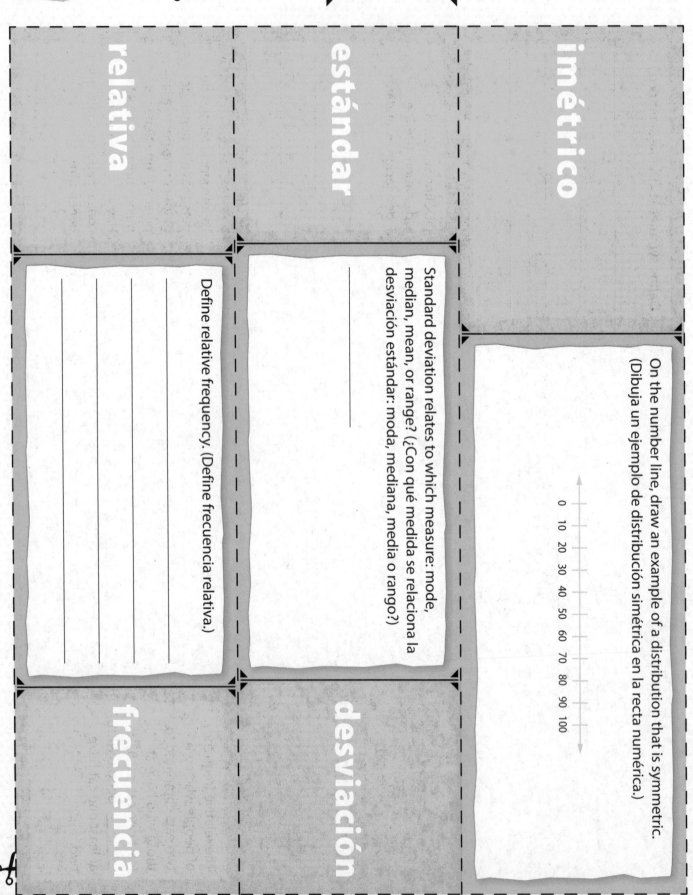

0
10
20
30
40
50
60
70
80
90
100

frecuencia

desviación

Dinah Zike's
Visual
Kinesthetic
Vocabulary®

✂ cut on all dashed lines   📷 fold on all solid lines

bivariate data

univariante

Circle the correct word to complete the sentence.

Comparing the numbers of pools built in different months is an example of (univariate, bivariate) data.

(Encierra en un círculo la palabra que completa correctamente la oración.

Comparar la cantidad de piscinas construidas en distintos meses del año es un ejemplo de análisis de datos (univariados, bivariados).)

cut on all dashed lines

fold on all solid lines

datos bivariantes

univariate

Circle the correct word to complete the sentence.

The numbers of chicks hatched by twelve different chickens is an example of (univariate, bivariate) data.

(Encierra en un círculo la palabra que completa correctamente la oración.

La cantidad de pollos incubados por doce gallinas distintas es un ejemplo de datos (univariados, bivariados).)